WARNING:

EXTRATERRESTRIALS HAVE COME TO TEST YOUR INTELLIGENCE WITH MIND-STRETCHING BRAINTEASERS

Without moving their mouths, the aliens before me communicated that they had a test to give me, a test for humanity. Were they trying to assess humanity's intelligence before entering into a partnership with us, I wondered? Were we simply guinea pigs in an experiment we could not fully understand? And then the most important question hit me. Was I up to the challenge?

—From *The Alien IQ Test*

ARE YOU UP TO THE CHALLENGE?

THE ALIEN IQ TEST

Previous Works by Clifford A. Pickover

Black Holes: A Traveler's Guide

Chaos in Wonderland: Visual Adventures in a Fractal World

Computers, Fractals, Chaos (Japanese)

Computers, Pattern, Chaos, and Beauty

Computers and the Imagination

Future Health: Computers and Medicine in the 21st Century

Fractal Horizons: The Future Use of Fractals

Frontiers of Scientific Visualization (with Stu Tewksbury)

Keys to Infinity

The Loom of God

Mazes for the Mind: Computers and the Unexpected

Mit den Augen des Computers

The Pattern Book: Fractals, Art, and Nature

Spider Legs (with Piers Anthony)

Spiral Symmetry (with Istvan Hargittai)

Visions of the Future:
Art, Technology, and Computing in the 21st Century

Visualizing Biological Information

THE ALIEN IQ TEST

CLIFFORD A. PICKOVER

BasicBooks
A Division of HarperCollins*Publishers*

Published by BasicBooks,
A Division of HarperCollins Publishers Inc.

Designed by Elliott Beard.

Library of Congress Cataloging-in-Publication Data
Pickover, Clifford A.
The alien IQ test / by Clifford A. Pickover. — 1st ed.
p. cm.
Includes bibliographical references.
ISBN 0-465-00110-6
1. Psychological recreations. 2. Puzzles. 3. Life on the other planets—Miscellanea. I. Title.
GV1507.P9P53 1997
793.73—dc21 96-37050
CIP

97 98 99 00 01 ❖/RRD 10 9 8 7 6 5 4 3 2 1

To the chordates, to Amphioxus

This book is also dedicated to those:
23815 251295225 208525 81225 25514 1242132054, 1144 2015
20815195 23815 81225 25514 1242132054. 1121915
2015 1291292021, 129129208, 1211391, 1144
2085 131185—2085 891920151893112
1618545351919151819 156 2085
82516141715793 “15124 817”
16851415135141514.

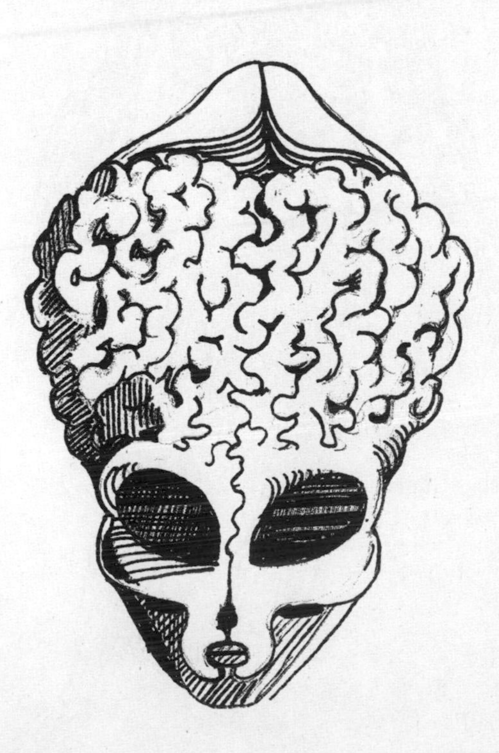

In my dreams
Amphioxus screams.

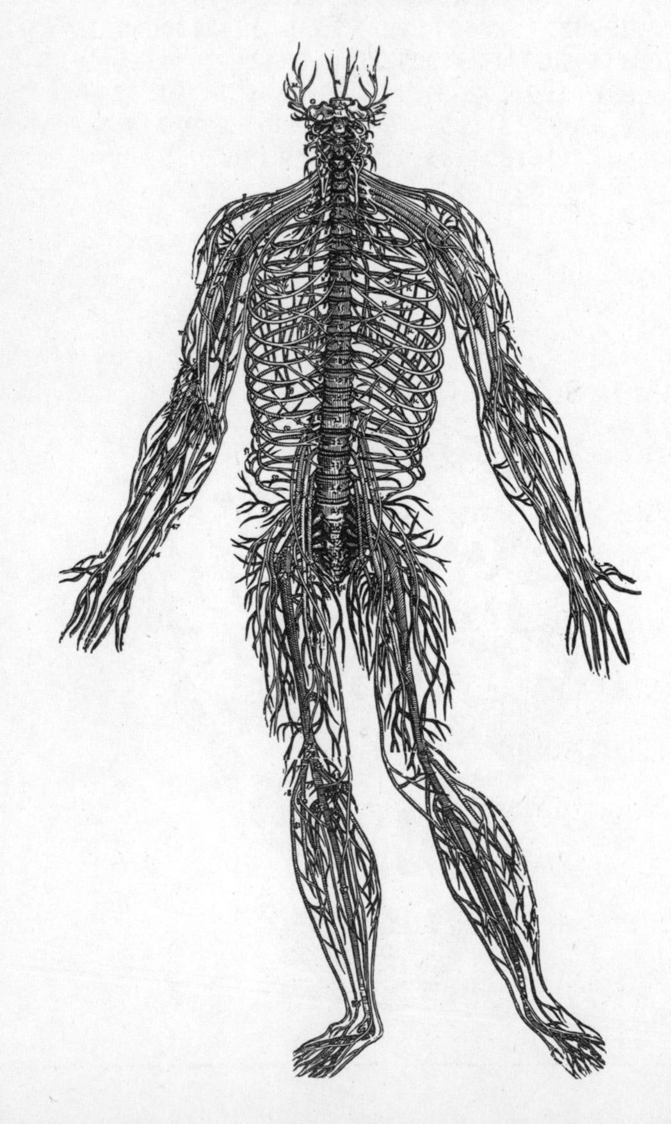

CONTENTS

ACKNOWLEDGMENTS xiii

PREFACE xv

1 Who This Book Is For 1

2 Symbols and Difficulty Levels 4

3 Alien Tiles 7

4 Alien Sperm 8

5 Alien Ellipses 10

6 Alien Repeats 12

7 Alien Matrix 14

8 Internal Organs 15

9 Alien Dissection 16

10 Alien Addition 18

11 Hyperdimensional Sz'kwa 20

12 Alien Spiral 22

13 Survival on Arcturus 23

14 Alien Medallion with Lights 24

15 The Omega Prism 25
16 Alien Worm 28
17 Alien Homoptera 30
18 Star Chart 32
19 Alien Spores 1 34
20 Alien Spores 2 36
21 Alien Spores 3 37
22 Alien Spores 4 38
23 Alien Spores 5 39
24 Rubik's Tesseract 40
25 Animal Eye 43
26 Cosmic Rosetta Stone 44
27 Alien Ants in Hyperspace 46
28 A Severed Human Finger 48
29 The Antikythera Mechanism 49
30 Alien Scrambling 50
31 Alien Aesthetics 51
32 Alien Knowledge and Talent 52
33 The Sagittarius Maneuver 54
34 Siriusian Geometry 56
35 Human Brains in a Jar 57
36 Human Belief Structure 58

37 Contact from the Pleiades 59

38 The Elk Hunter's Abduction 60

39 Loss of Scientific Knowledge 61

40 Aliens and Sprinklers 62

41 Unanswered Questions 64

42 Moral and Emotional Choices of Humans 67

43 Coded Transmission 70

SOLUTIONS 73

FOR FURTHER READING 99

THE MISSED MILLENIUM 103

ABOUT THE AUTHOR 105

"No live organism can continue for long to exist sanely under conditions of absolute reality. Even larks and katydids are supposed, by some, to dream."

SHIRLEY JACKSON

ACKNOWLEDGMENTS

> The dream might have been more than a dream. It was as if a door in the wall of reality had come ajar . . . and now all sorts of unwelcome things were flying through.
>
> STEPHEN KING, *Insomnia*

> We might find ourselves in alien test tubes set up to investigate us as we do guinea pigs. If the space phone rings, for God's sake, let us not answer it.
>
> ZDENEK KOPAL,
> Czech-British astronomer

> *Τα μικρα πρασινα πλασματα με επισκευτικαν την νυχτα*
>
> CHARON CHRISTIDIS

> I am a camera with its shutter open. . . .
>
> CHRISTOPHER ISHERWOOD

Abductee Clay Fried from Westchester, New York, drew from memory the alien images on pages vii, xiv, xvii, xix, 11, 27, 33, 42, 47, 58, and 63.

The alien creatures on pages xii, xx, 2, 102, and 104 are from Marc Christopher Williams (e-mail: pozzo@texas.net).

"The Missed Millennium" is a fragment from a longer poem by Keith Allen Daniels.

The mathematics of Rubik's tesseract were discussed in: Velleman, D. (1992) Rubik's tesseract, *Mathematics Magazine*, February 65(1): 27–36. I thank Kermit Hummel and Susan Rabiner of Basic Books for helping to bring this project to fruition and for their continued support and encouragement.

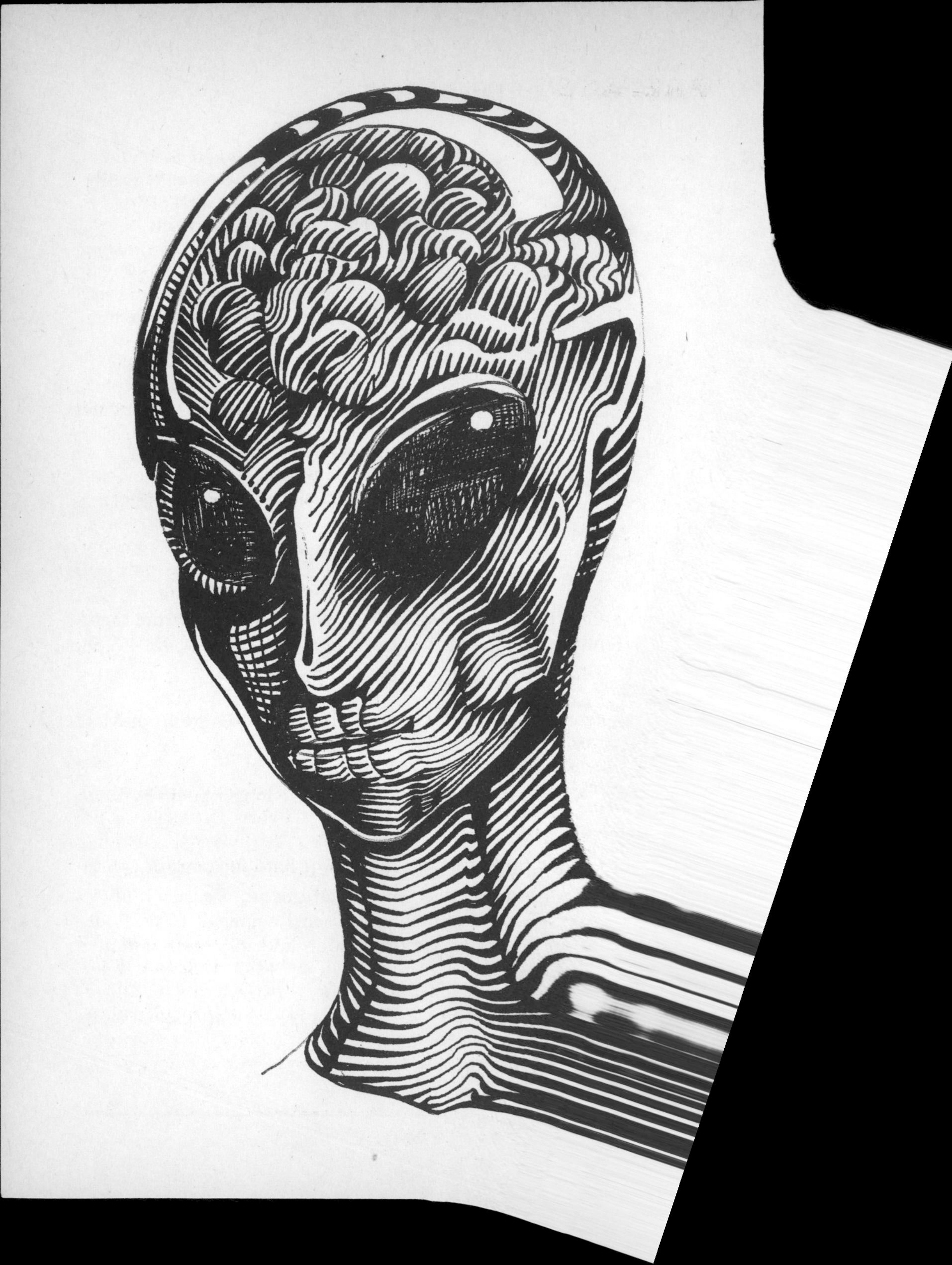

ACKNOWLEDGMENTS

> The dream might have been more than a dream. It was as if a door in the wall of reality had come ajar . . . and now all sorts of unwelcome things were flying through.
>
> STEPHEN KING, *Insomnia*

> We might find ourselves in alien test tubes set up to investigate us as we do guinea pigs. If the space phone rings, for God's sake, let us not answer it.
>
> ZDENEK KOPAL,
> Czech-British astronomer

> *Τα μικρα πρασινα πλασματα με επισκευτικαν την νυχτα*
>
> CHARON CHRISTIDIS

> I am a camera with its shutter open. . . .
>
> CHRISTOPHER ISHERWOOD

Abductee Clay Fried from Westchester, New York, drew from memory the alien images on pages vii, xiv, xvii, xix, 11, 27, 33, 42, 47, 58, and 63.

The alien creatures on pages xii, xx, 2, 102, and 104 are from Marc Christopher Williams (e-mail: pozzo@texas.net).

"The Missed Millennium" is a fragment from a longer poem by Keith Allen Daniels.

The mathematics of Rubik's tesseract were discussed in: Velleman, D. (1992) Rubik's tesseract, *Mathematics Magazine*, February 65(1): 27–36. I thank Kermit Hummel and Susan Rabiner of Basic Books for helping to bring this project to fruition and for their continued support and encouragement.

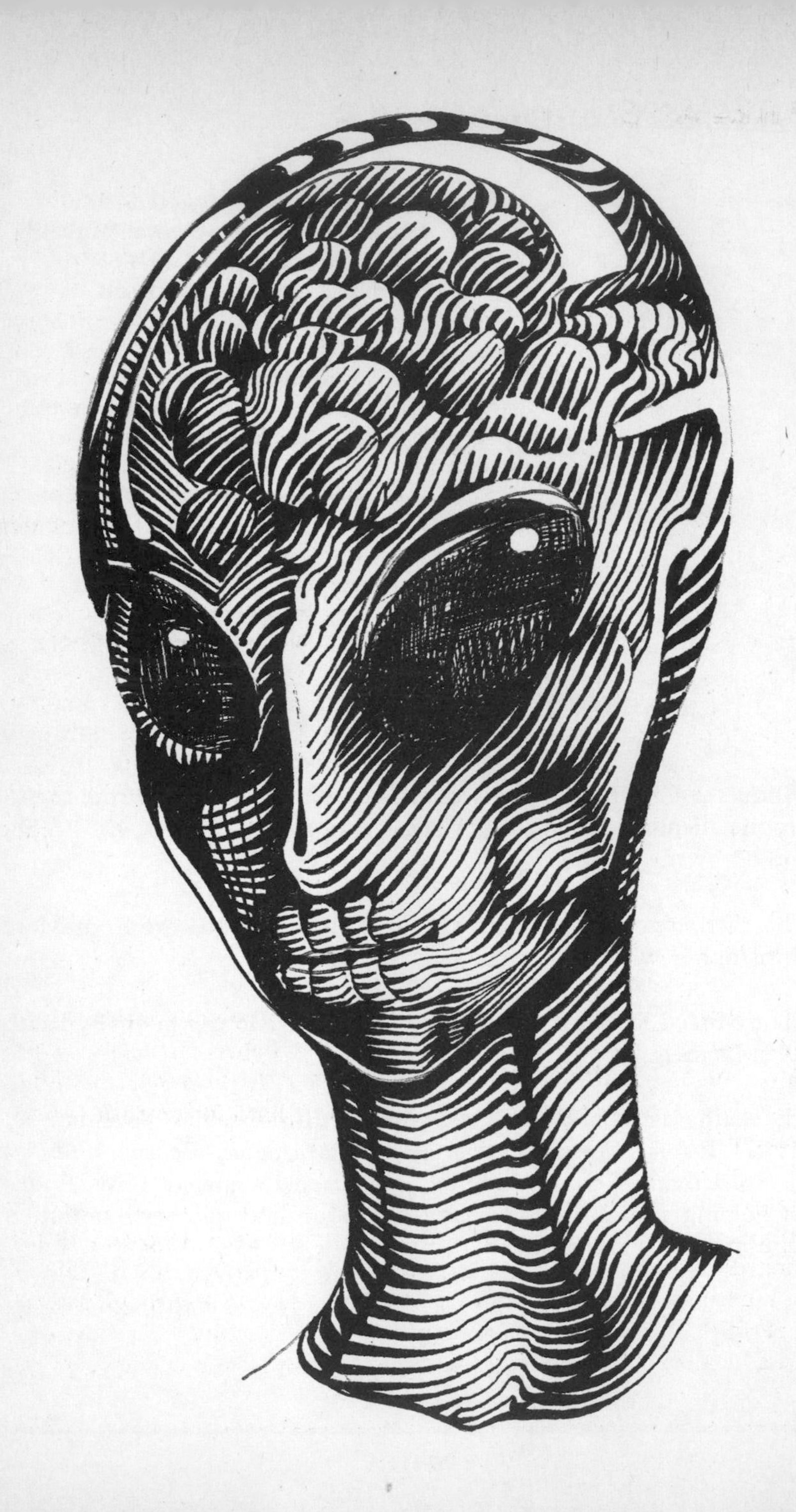

PREFACE

Our normal waking consciousness is but one special type of consciousness, whilst all about it, parted from it by the filmiest of screens, there lie potential forms of consciousness entirely different.

No account of the universe in its totality can be final which leaves these other forms of consciousness quite disregarded. They may determine attitudes though they cannot furnish formulas, and open a region though they fail to give a map.

WILLIAM JAMES

It may well be doubted whether human ingenuity can construct an enigma of the kind which human ingenuity may not, by proper application resolve.

EDGAR ALLAN POE

The soul of man was made to walk the skies.

EDWARD YOUNG

NIGHT VISION

The strange looking puzzles in this book come from another world—a world where small gray beings visit us only at night. Their bodies are spindly and thin. Their heads are huge, their eyes shiny, oval, slanted, and black. I know these ideas are going to be hard for most of you to swallow.

For years I have been getting ideas for books, art, and research through dreams or during the twilight realm between sleep and awakening. Probably many of you also get creative sparks in the middle of the night or while letting your mind drift as you walk through forests or perform some innocuous or relaxing task. To enhance my dreams and add more visual vitality, I practice various exercises before going to sleep. These mental exercises include rotating imaginary objects, and looking at imaginary lights and mirrors—all in the hypnagogic realm at the edge of sleep.

My interest in alien assessment of human intelligence started one night when I was researching the increasing world-wide interest in alien abduction. In the classic UFO abduction scenarios, abductees tell us that they experience bizarre dreams, memory flashes, or even physical symptoms weeks after an alien encounter. Some abductees undergo hypnosis in an attempt to recover memories of events that occurred during unexplained lapses of time. These lapses are sometimes called "missing time" when events that seemed to last a few minute's duration actually took hours.

Under hypnosis, abductees reveal that they have been led, sometimes floated, into disk-shaped crafts by aliens. In these ships, abductees undergo a medical examination of some kind. So far, there is not sufficient hard evidence that alien abduction exists outside the mind of the abductee.

AMPHIOXUS OVERDRIVE

As part of my research into the alien abduction phenomena for a chapter in my book *Strange Brains and Genius,* I surrounded myself with an array of books and articles: Whitley Streiber's *Communion* and *Transformation,* John Mack's *Abduction: Human Encounters with Aliens* (Revised Edition), Susan Blackmore's "Alien Abduction," Philip Klass' *UFO Abductions*, Budd Hopkins' *Intruders*, Eve LePlante's *Siezed*, and C. D. B. Bryan's *Close Encounters of the Fourth Kind: Alien Abduction, UFOs, and the Conference at M.I.T.*

After reading from about 10:00 to 11:30 at night, I put the books on the night table.

I was asleep for what seemed like a few hours when something peculiar happened. Nothing in my scientific intuition, nothing in my 15 years of gaining nocturnal creative insight prepared me for what stepped into my bedroom. The vertical blinds began to move. My heart seemed to stop. What could be moving the blinds? I turned on the light by my bed.

My incredulity grew as two beings floated through the screen covering the open window. When standing on the carpet, their heads were just slightly higher than the window sill. I noticed a slight odor of limes, or perhaps some fragrant flower. The two creatures were hairless and had no ears. Their huge black eyes had no whites or pupils—just like the descriptions in the book I had been reading.

Alright, this is just a dream, I said to myself. Go along with it. Maybe something interesting would happen. They moved a little closer to me, and I found I could only move my eyes to watch them. The creatures had thin limbs, with large, bulbous foreheads dominated by tear-shaped eyes. Their noses and mouths were small. Their skin was smooth with no distinguishing marks such as pores. They were about four feet tall, with the tiniest of chins. Their wet eyes never blinked or moved in their sockets. Their chests had no nipples.

Without moving their mouths, the aliens somehow communicated that they had a test to give me, a test for humanity. I didn't understand their motives. Were they trying to assess humanity's intelligence before entering into a partnership with humankind? Were we simply guinea pigs in an experiment we could not fully understand?

They started by gesturing to plaques filled with symbols. When they pointed at the plaques, I noticed that they had four boneless fingers and no thumbs. Before I could give the crazy experience much more thought, another being came through the window. It never came very close, but I could tell that the small gray beings were aware of its presence because they momentarily turned their heads to it.

I clenched my fists. If the "small grays" were scary, at least their image was so much a part of our culture that my reaction was not one of utter horror. But the new creature was something different. On each side of its long body rippled powerful longitudinal muscles divided into V-shaped segments. I could not see the end of its body, because only its tentacled head and a portion of its body protruded through my window. Having majored in biology at college, I recognized the floating creature in an instant. It was an amphioxus, also known as a lancelet.

Lancelet is the common name for about 25 species of simple marine animals, transitional between invertebrates (animals without backbones) and vertebrates (animals, like humans, with backbones). They have a stiff dorsal rod (notochord) but no vertebrae and no heart. Around their mouths, cirri and tentacles move like a bag of nervous worms. We share the same ancestry as amphioxus as evidenced by our embryonic tongue-barred gill-slits.

A fishy odor began to fill the air in my bedroom, quickly followed by the smell of absinthe. What was this thing doing in my room? What was the creature perceiving about me? Billions of years ago, the genetic code of primitive cells was passed from generation to generation and finally to multicellular organisms, then to invertebrates, and then to vertebrates around 600 million years ago. Because amphioxus are transitional forms between invertebrates and vertebrates like ourselves, the amphioxus can be thought of as the evolutionary gateway to humanity, to more developed brains, and to consciousness.

Amphioxus, the gate, born millions of years ago . . . Why do humans exist? The answer quite simply is because amphioxus and its kin were born eons ego and lived to survive.

My dream of the small gray aliens ended soon after the appearance of the amphioxus. As my bedroom returned to normalcy, I felt relieved but also somewhat excited. All was quiet. I had the feeling that I was on a beach at low tide, after the waters had withdrawn and left behind a quiet, peaceful beach.

For days after this experience, weird-looking alien test symbols swarmed behind my eyelids as I tried to sleep. I'll explain the symbols and their use in coming chapters.

Do you dare read further?

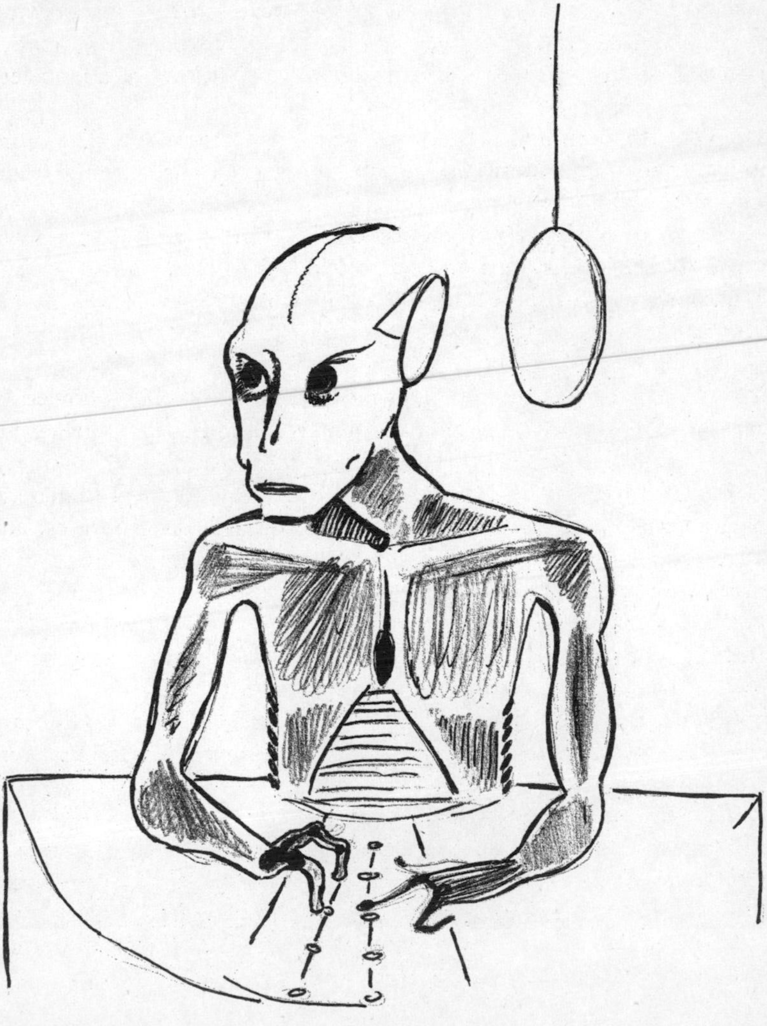

1 WHO THIS BOOK IS FOR

> All beings sufficiently intelligent for interstellar communication must have a mathematics. . . . Yet their higher mathematics, their logic, their way of representing atomic structure, may differ radically from our own.
>
> WALTER SULLIVAN, *We Are Not Alone*

> You're drifting into deep water, and there are things swimming around in the undertow you can't even conceive of.
>
> STEPHEN KING, *Insomnia*

> These are much deeper waters than I had thought.
>
> SIR ARTHUR CONAN DOYLE, *The Memoirs of Sherlock Holmes*

This book is for anyone who wants to enter new mental worlds: puzzle-solvers including students, behaviorists, psychologists, educators, cryptographers, computer programmers, CIA agents, and psychiatrists studying alien abductees. If you are a teacher, you may want to use the cryptic designs to stimulate students. Have them design their own puzzles similar to the ones in this book. Computer programmers may want to create or solve similar puzzles, although a computer is definitely not necessary to attack and solve the problems in this book.

Some of you may question my use of the term "IQ" in this book's title to indicate intelligence. I use the term loosely to indicate the degree to which humans can make abstractions, learn, deal with novel situations, and understand mathematics. In my mind, the term "high IQ" implies high intelligence involving the ability to reason, solve problems, plan, and comprehend complex ideas.

Critics have argued that IQ tests favor groups from more affluent backgrounds, and consequently other intelligence tests are being considered. Some have even tried to measure IQ directly using electroencephalographs of the brain's electrical activity, but this approach is still experimental. Will critics of standard IQ tests consider *The Alien IQ Test* less culturally biased than traditional tests? Some researchers, such as Howard Gardner at Harvard University, suggest that there are really seven broad forms of intelligence: linguistic, spatial, logical-mathematical, musical, bodily kinesthetic, intrapersonal, and interpersonal. Obviously, standard IQ tests do not test for intelli-

gence in all these areas. *The Alien IQ Test* perhaps stimulates a broader range of mental facilities than traditional tests and also incorporates some "human literacy" problems. For example, if aliens visited us and asked, "What does it mean to be human? What do human beings think about?" how much could you tell them?

Think of this book as a mental ocean. As you surf, each class of puzzle exercises a different style of thinking. Be careful not to get caught in the undertow.

Enjoy the cryptic waters. Take deep breaths. Don't stay in too long. Let your mind rest between sections. Many books have been written on the advantages of developing different thinking styles; for example, see Howard Gardner's *Frames of Mind*, or any of my previous books such as *Mazes for the Mind*.

My intention is to stretch your mental powers to extremes.

2 SYMBOLS AND DIFFICULTY LEVELS

> If an alien comes to you and asks, "What is the most important question we can ask humanity and what is the best possible answer you can give?," the safest reply is, "You have just asked the most important question you can ask humanity, and I'm giving you the best possible answer."
>
> CLIFFORD A. PICKOVER

> If we wish to understand the nature of the Universe we have an inner hidden advantage: we are ourselves little portions of the universe and so carry the answer within us.
>
> JACQUES BOIVIN, *The Single Heart Field Theory*

SYMBOLS

I pondered my dream for a few days. The small grey beings had shown me puzzles (intelligence tests?) printed on thin plaques filled with an array of exotic looking symbols. I gradually grew certain that each symbol stood for a number and that some of them were arranged in a maze for me to solve. As in most dreams, I found it difficult to carefully gaze at the printed text and see sharp, easily remembered patterns. Nevertheless, I roughly recalled three of the symbols, because I wrote them down immediately upon wakening.

One symbol resembled a human figure with spots floating around its head. This was the first symbol the aliens pointed to with their thin fingers; and I felt this symbol must stand for the number "1." Another resembled a symmetrical chandelier or menorah with several dangling balls. The only thing I recall from the dream is that this was not an ordinary number, but rather it was probabilistic in nature. It could represent one number or another number, as useful to solve a particular problem.

The only other "number" I recall looked like a comet streaking downward. Although I cannot remember the remainder of their symbols, I have chosen a set that gives a feeling as to the general shape and structure of their numerals. I have also given each symbol a name so that we can conveniently talk about them. The table below lists the numeric values and arbitrary names for each symbol. As I said, in some of the puzzles, the menorah symbol is quite strange-probabilistic. It can be either a 5 or a 6.

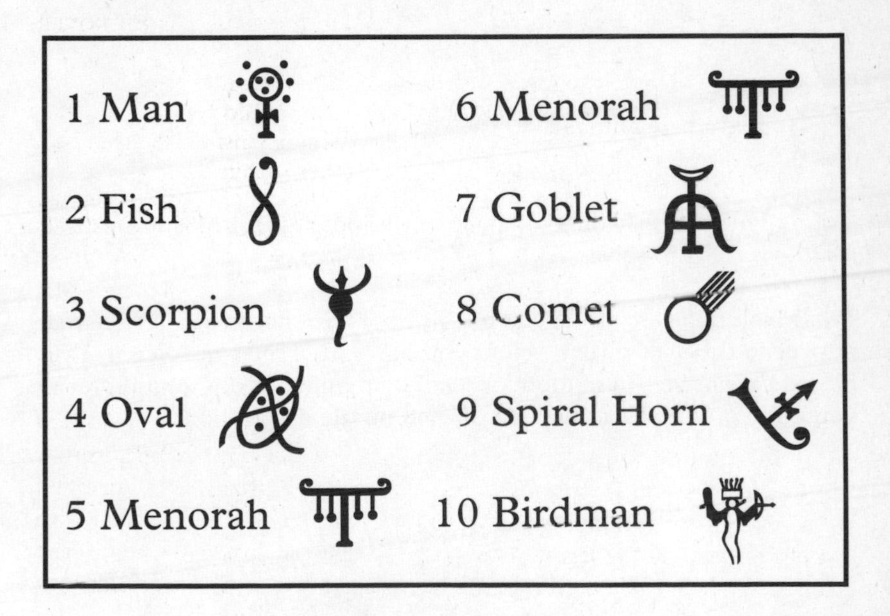

I've done my best to capture the spirit of the problems shown to me by the gray beings. Don't feel bad if the puzzles are too difficult to solve on your own. Find some friends and work in teams. If the puzzles were an IQ test for humanity from outer space, then perhaps the aliens wouldn't mind us trying to collectively solve the problems.

DIFFICULTY LEVELS AND SOLUTIONS

To help you assess your level of performance during your journey through this book, I have assigned difficulty ratings to the various puzzles:

 Challenging

 Very Challenging

 Extremely Difficult

 Outrageously difficult: probably impossible for most terrestrial readers to solve

Puzzle solutions are at the end of this book. So that you don't see the answer to the next puzzle before you have attempted to solve it, I've placed the answers in random order. The number of the solution for a particular problem is at the end of each puzzle description.

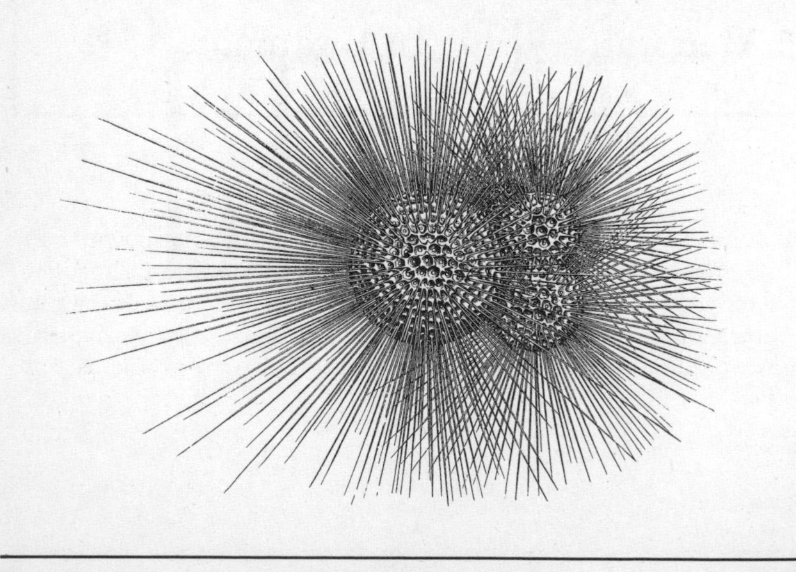

3 ALIEN TILES

> The hardest thing to imagine is the space between the stars.
>
> A. HERBERT

> Dark with excessive bright.
>
> JOHN MILTON, *Paradise Lost*

The first puzzle in this book consists of five "tiles" each containing two alien symbols:

You are to find a pair of symbols to complete the sequence, by choosing the correct tile from among the following five possible solutions:

Hint: You need not assign numerical values to each symbol to determine which pair completes the set.

Difficulty rating: 

Answer: Ans10

4 ALIEN SPERM

One of the purposes for which UFOs travel to Earth is to abduct humans to help aliens produce other Beings. It is not a program of reproduction but one of production. They are not here to help us. They have their own agenda, and we are not allowed to know its full parameters. . . . The focus of the abduction is the production of children.

DAVID JACOBS,
Secret Life:
Firsthand Accounts of UFO Abductions

Shown here are eight patterns, each containing three alien sperm.

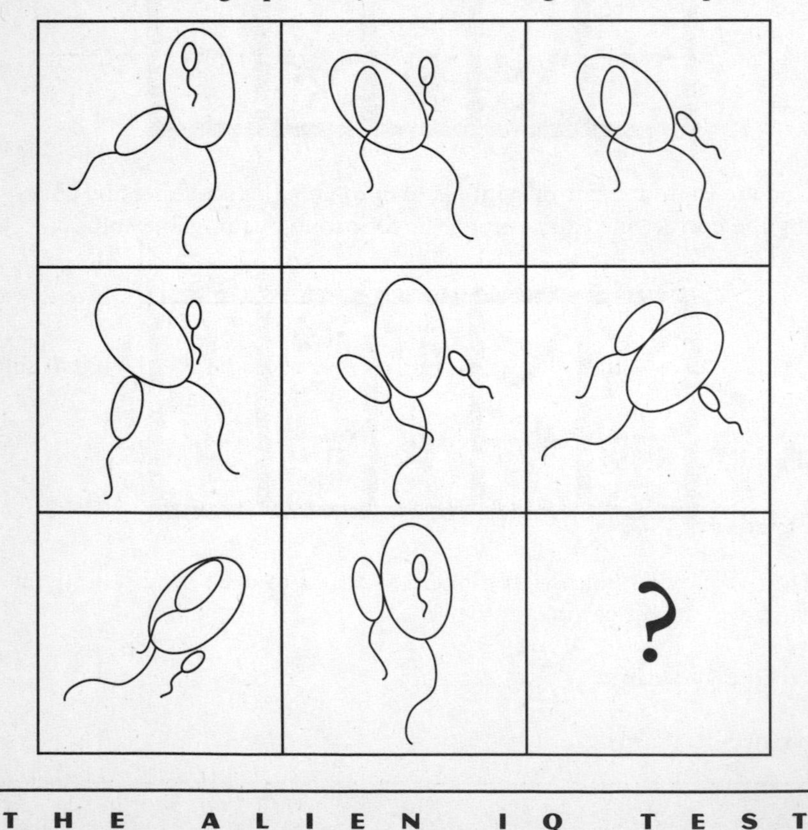

The question mark at the bottom right is to be replaced with one of the nine patterns below:

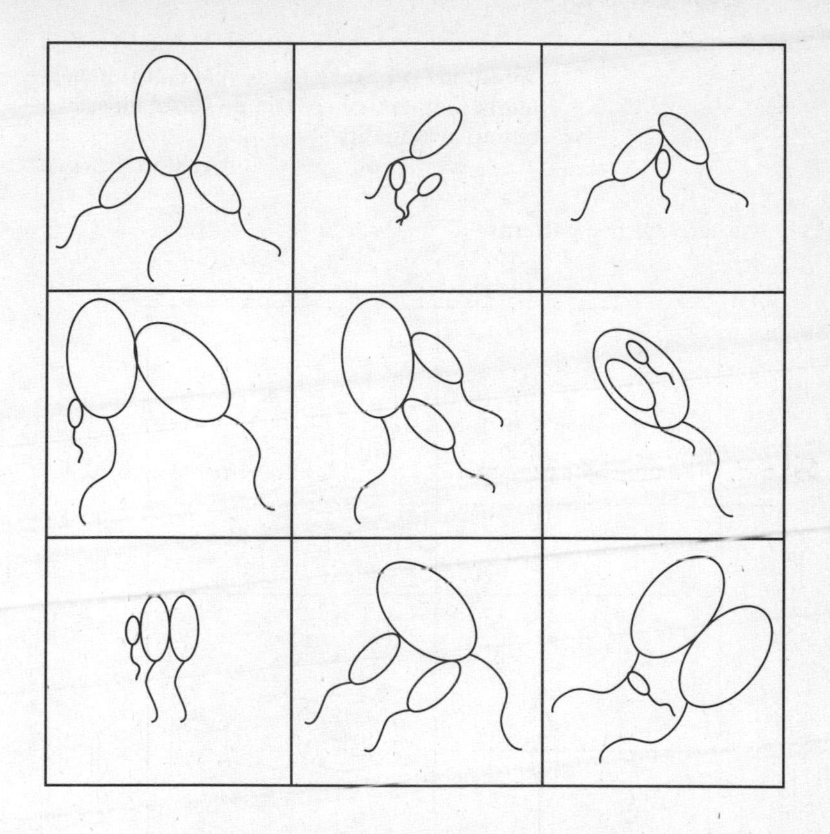

Which of the nine patterns do you choose to complete the initial set, and why?

Difficulty rating: 

Answer: **Ans46**

5 ALIEN ELLIPSES

> We should take care not to make the intellect our god; it has, of course, powerful muscles, but no personality.
>
> ALBERT EINSTEIN

Shown here are five patterns.

Replace the missing pattern with one of the four patterns below:

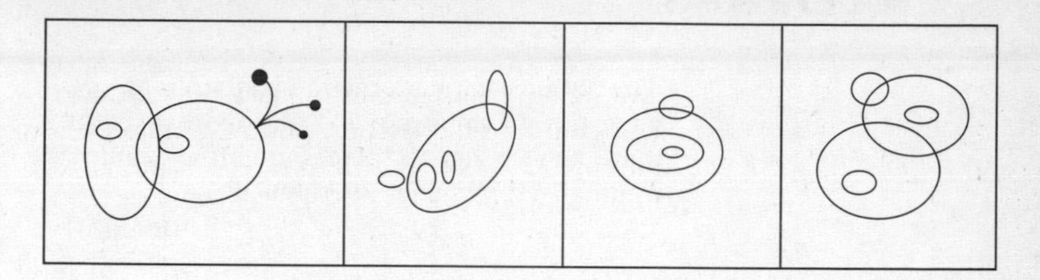

Which of the four patterns do you choose, and why?

Difficulty rating:

Answer: Ans16

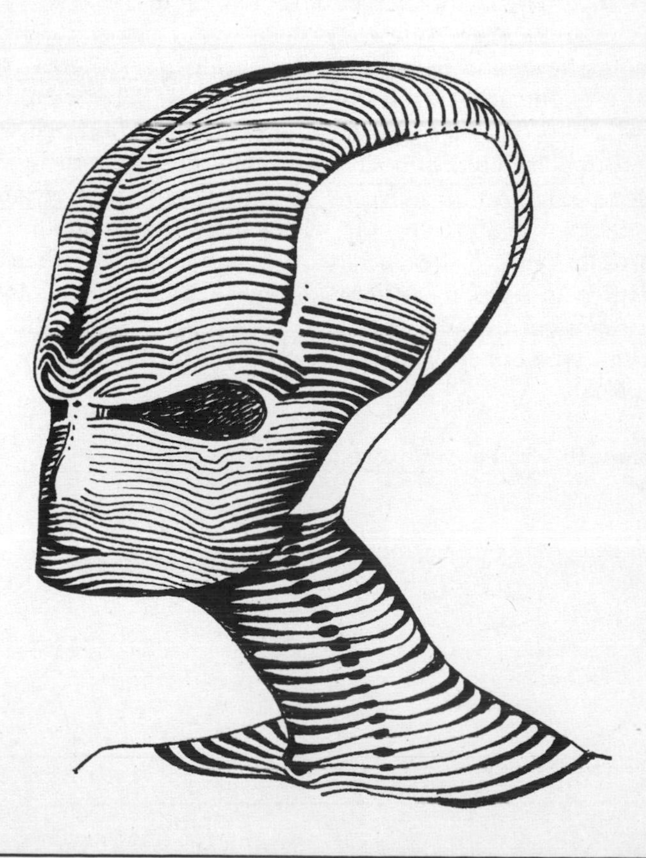

6 ALIEN REPEATS

> The abduction phenomenon is, therefore, of great clinical importance if for no other reason than the fact that abductees are often deeply traumatized by their experiences.
>
> JOHN MACK

Hiding within the following intricate array of symbols is a smaller rectangular group of symbols that appears more than once. What is the largest block of repeated symbols you can find? How long did it take you to find the repeated group? Would you be willing to wager $1000 that there is not a larger block to be found?

Warning: Several colleagues have developed eye stress when they stared too long at the arrays of alien symbols. What would happen if the two cultures produced a form of communication that was painful to one another? This is not an outrageous idea. Reports in an October 1989 issue of *Archives of Neurology* demonstrate that certain patterns help distinguish those people who suffer from migraine headaches from other types of headaches. Migraine sufferers, when presented with the pain-inducing patterns, found them extremely objectionable and attempted to avert their gaze, while people who did not suffer from this type of headache had relatively little difficulty looking at the patterns.

Be calm. Proceed at your own risk.

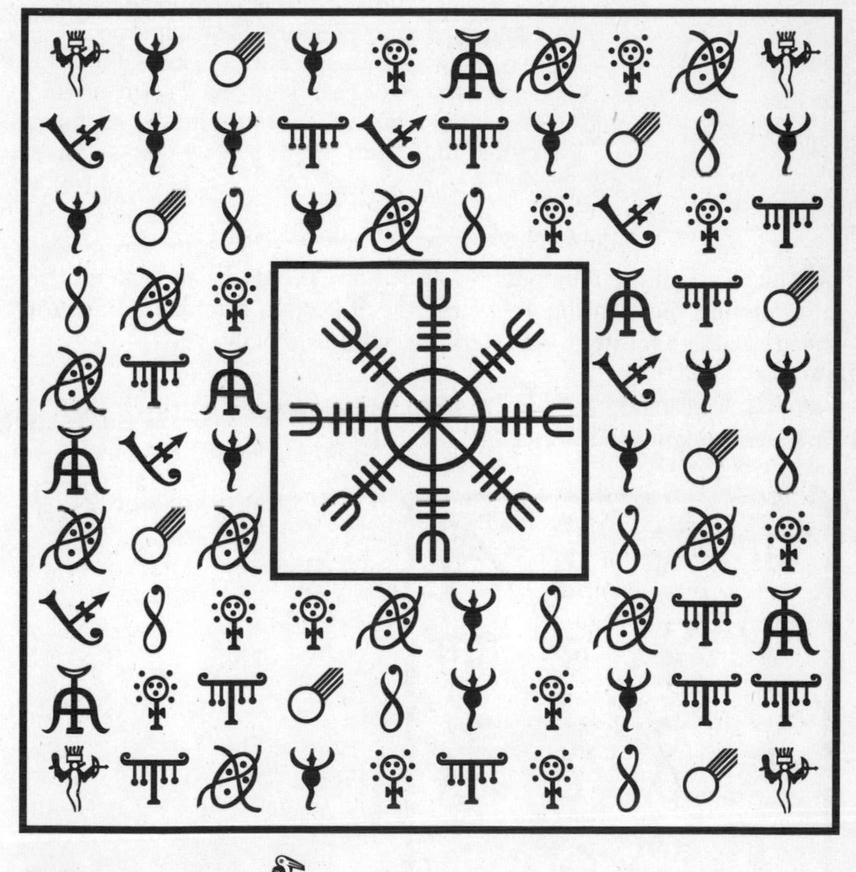

Difficulty rating:

Answer: Ans15

7 ALIEN MATRIX

A primitive creature may take apart an internal-combustion engine to study it but still never understand how it works—because its secret lies external to it, in the principle that explosions exert pressure. Hence it is no mystery why scientists haven't grasped the brain: they have been studying it solely on its own terms, much the way a primitive creature studies the engine.

TOM SALES

Which one of the four symbols at bottom should be used to replace the missing space in the 4 × 4 matrix of framed symbols? Hint: You need to assign numeric values to the symbols to solve this.

What is the logic you used to solve this problem? Is there another logic you might use to solve it differently?

Difficulty rating:

Answer: **Ans29**

8 INTERNAL ORGANS

Why does there seem to be something inhuman about regarding human beings like roses and refusing to make any distinction between the inside of their bodies and the outside?

YUKIO MISHIMA

Aliens wish to assess how well we know ourselves. They have therefore borrowed the organ system shown here from a Kansas City mill worker and asked us to identify it. Can you identify this body part?

Difficulty rating: 

Answer: **Ans42**

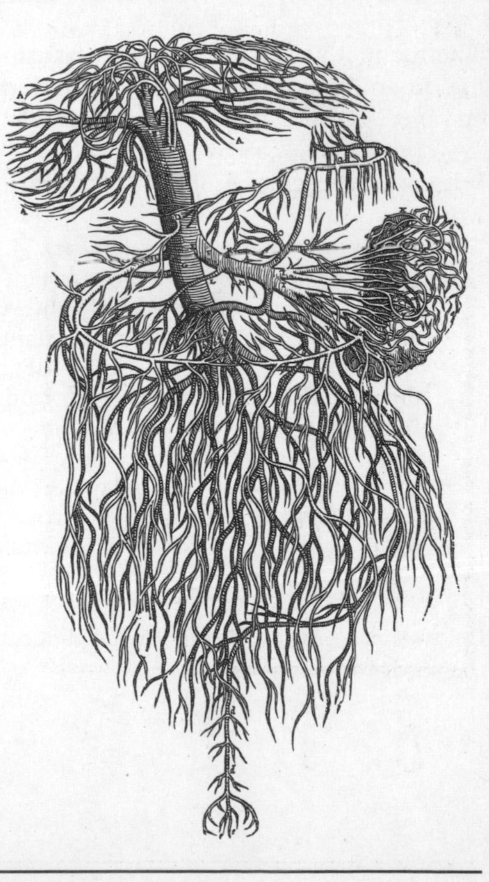

9 ALIEN DISSECTION

> The life of every man is a diary in which he means to write one story and writes another.
>
> SIR J. M. BARRIE

A Metro-North commuter train is traveling from Croton-Harmon to Ossining, New York, when suddenly the train conductor sees a dark monolith protruding from the tracks. The conductor immediately slams on the train's brakes, bringing the train to a grinding halt inches from the monolith. The conductor exits the train and approaches the dark shape.

The monolith sits in a capsule of translucent white light as bright as fog on an Autumn morning. The light transforms the monolith into an object of great dignity and unbearable beauty.

Within a deep crevice of the monolith are dozens of cards with the words:

> We come from a far-away star system and wish to assess your intelligence. Create two areas, both of exactly the same size and shape, so that both areas contain equal numbers of each symbol. You can use a pencil to define the areas, but all the lines you draw must be either horizontal or vertical. (You may think of this as cutting a square cake into two identically shaped pieces.) If you and your train passengers are unable to complete the task within a day, we will transport the train and its occupants to our world for further study. Please do not attempt to leave the vicinity of the train. Non dormivo. Stavo soltanto riposando gli occhi.

Below these words is the following diagram. The conductor hands out the cards to the passengers.

Hours have passed, and no one has been able to solve the problem. Can you help?

Difficulty rating: 

Answer: **Ans33**

10 ALIEN ADDITION

I looked round the trees. The thin net of reality. These trees, this sun. I was infinitely far from home. The profoundest distances are never geographical.

JOHN FOWLES, *The Magus*

The Metro-North commuter train in the previous puzzle is continuing its journey from Marble Hill to Grand Central Station, New York, when suddenly the train conductor hears a cry.

Without warning, dozens of small gray aliens enter the train and begin handing out pieces of parchment containing arrays of symbols.

Again, this is a test of some sort. Above the array of symbols are the following instructions:

> Starting at any symbol touching the outer edge of the outer frame (for example the scorpion in the upper left), you must travel left, right, up, or down through the grid of symbols. As you pass through the symbols, you add their values to a running total. Your goal is to start at an outer-edge symbol and finish at an outer-edge symbol with a cumulative sum of 114. What is your path?

The conductor clenches his fists. "And if we refuse to solve the puzzle?"

One of the aliens takes a step closer. "Then the train will never reach its destination. It will ride on an infinitely long track, forever. In addition, we will place the train in a parallel universe where history is slightly different from the one you know. Perhaps a universe where the Greek civilization was never destroyed. Or where Kennedy was never assassinated. Or where Hitler conquered the world."

Difficulty rating: 

Answer: **Ans2**

11 HYPERDIMENSIONAL SZ'KWA

> Imagine an ant finding its path suddenly blocked by a discarded Styrofoam cup. Even if the ant is intelligent, can it hope to understand what the cup is for and where it came from?
>
> CHARLES PLATT,
> *When You Can Live Twice as Long, What Will You Do?*

An alien challenges you to a simple-looking, competitive game. You hold 25 white crystals in your hand; the alien holds 25 black crystals. At the start of the game, the circular board shown here would have no crystals.

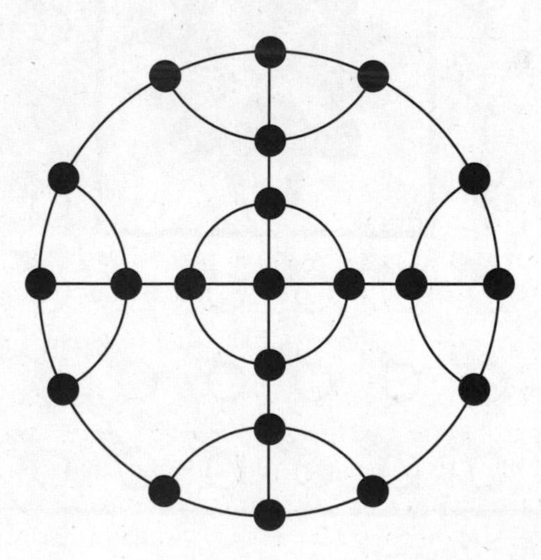

You and the alien take turns placing a crystal on the board, at the positions with black dots. The rules are as follows. If a player's crystal is completely surrounded by the opponent's crystals, it is captured. The following diagram shows the capture of a black piece (top), and the capture of two white pieces (bottom).

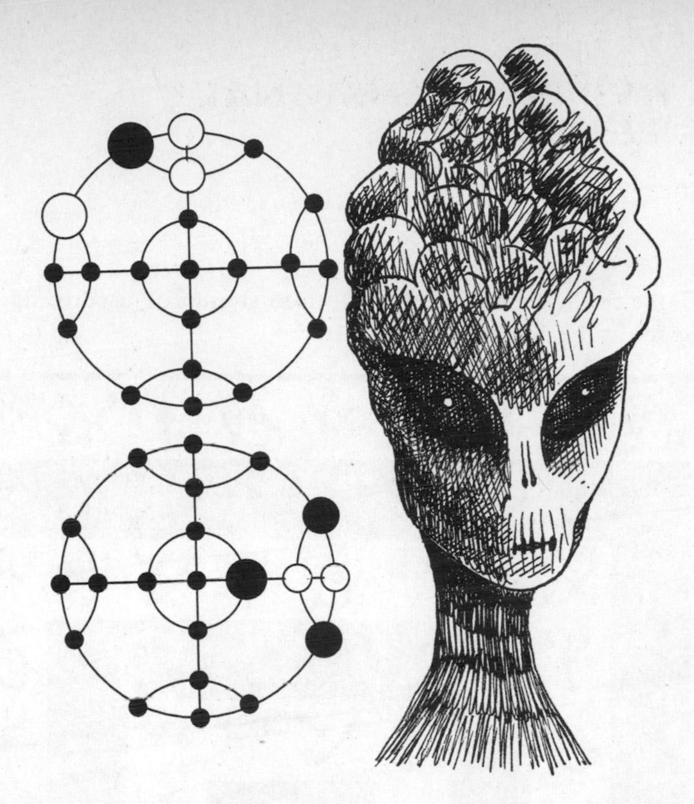

When a player has no crystals left to place on the circular board, or no empty sites on which to place a crystal without it being captured, the game ends. The winner is the player who holds the greatest number of crystals.

How many different arrangements of crystals on the playing board exist? Is it better to be the first player? Can you write a computer program that learns strategies by playing hundreds of games and observing its mistakes? Develop a multidimensional Sz'kwa game in which the center site on the Sz'kwa board connects center sites on adjacent boards. First try a game using just two connected boards, and then three. Generalize your discoveries to multiple connected boards.

Difficulty rating: 

12 ALIEN SPIRAL

Αυτο θα ειναι στην εζεταση?

Which one of the four symbols at bottom should be used to fill the empty space at the center of the spiral?

Difficulty rating: **Answer: Ans1**

13 SURVIVAL ON ARCTURUS

> The universe is not only stranger than we imagine; it is stranger than we can imagine.
>
> J. B. S. HALDANE

Aliens have abducted you and taken you to an Earth-like planet circling Arcturus in constellation Bootes (*boo-oh-teez*). Arcturus, a solitary yellow star like our sun, is 12 light years away from Earth. The planet is covered with a lush forests, and contains flora and fauna vaguely reminiscent of those found on a tropical island on Earth.

You must survive on the planet for one year before the aliens will bring you back to Earth. You may take one of the eight objects below:

1. a bottle of cologne
2. a pin
3. a pencil
4. a pen
5. a TV remote control
6. a Rubik's cube
7. a rodent
8. a book of matches

When I asked dozens of scientist-colleagues which object they would choose, there was one clear favorite. Which object do you think was chosen most frequently? Which do you choose and why?

Difficulty rating: 

Answer: **Ans49**

14 ALIEN MEDALLION WITH LIGHTS

She arrives home and goes through the rooms of her house looking for someone, looking out the windows. She feels a sense of urgency, a sense that someone is coming or that something is going to happen.

JOE NYMAN

The top four medallions contain numerous small lights (circles), some of which are turned off (blackened). Choose one of the two bottom medallions that belongs with the set of top four medallions.

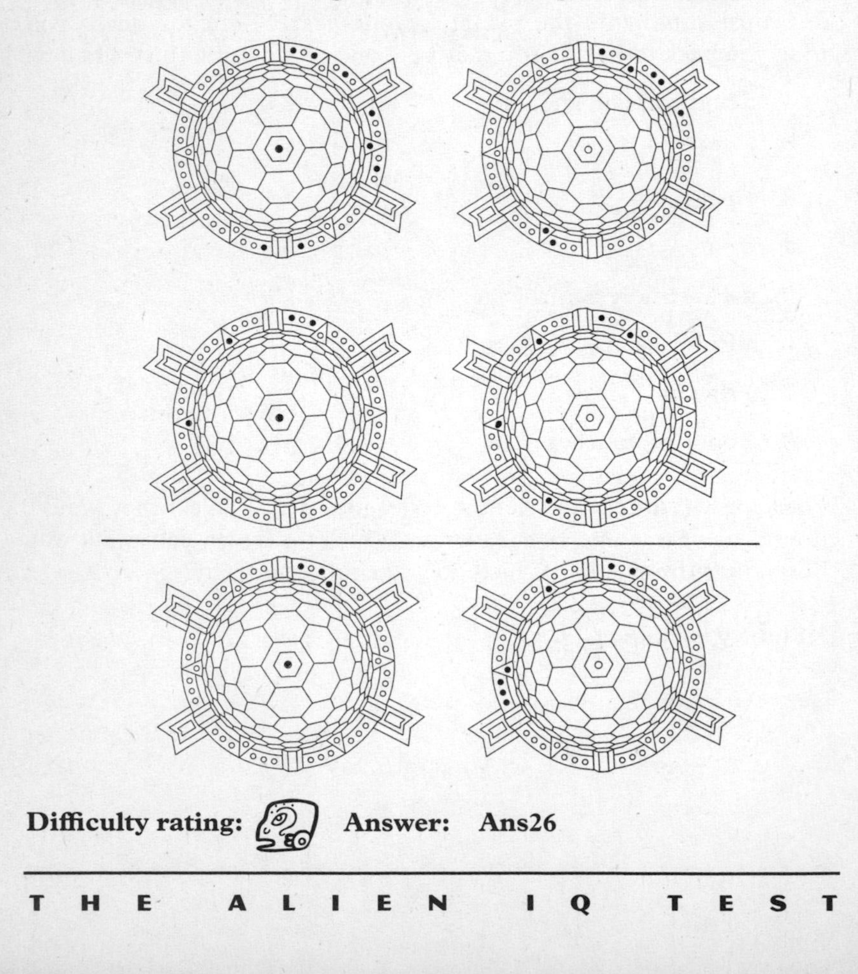

Difficulty rating: **Answer:** **Ans26**

15 THE OMEGA PRISM

They squatted among the dead flowers and macaws, and played with the pennants of my blood. Rachel fondled them, her blind eyes flickering rapidly, trying to read their mysterious codes, cryptic messages from another universe transmitted by the ticker-tape of my heart.

J. G. BALLARD,
The Unlimited Dream Company

I can conceive of no nightmare so terrifying as establishing communication with a so-called superior (or, if you wish, advanced) technology in outer space.

NOBEL LAUREATE GEORGE WALD,
Harvard University

On one cool night in November, a Kansas City mill worker sees a streaking across the sky. After a few minutes, there is a glowing in his corn field. Upon closer inspection, he finds an object resembling a Rubik's cube protruding from the ground.

When he picks up the crystalline object, he finds that it is nearly cubical, and its six faces are tiled in a colorful substance that luminesces. Each square tile appears to be about a millimeter (mm) in length. On the ground by the object is a note that reads:

> You hold in your hands an Omega Prism, a 230 mm × 231 mm × 232 mm brick whose faces are tiled with 1 mm × 1 mm squares. If you were to draw a straight line on the rectangular faces from one corner to another, on which face does the diagonal line cross the most tiles? Can you determine the number of tiles crossed for any face? To solve this puzzle, you are not permitted to trace a diagonal on a prism face and count the number of tiles crossed. We are watching. If you fail to solve the puzzle

within a week, we will colonize the Earth and use humans as food for further thought.

The mill worker stares at the Omega Prism for several minutes, clenches his fists, and throws the prism to the ground. Even if he were allowed to trace the diagonal with a marker, the colors are blinking so rapidly that it would be nearly impossible for him to count the crossed tiles. A wind begins to blow through his field—a cold wind that sounds like the chanting of monks.

Simultaneously, Omega Prisms land near corn fields in Lexington, Kentucky; Yorktown Heights, New York; Uithoorn, Netherlands; Sydney, Australia; Fukushima, Japan; Torrejon, Spain; Winchester, United Kingdom; Armonk, New York; and Raleigh, North Carolina.

Can you help save the Earth? Given just the side lengths of Omega Prisms, can you determine the number of square tiles through which a diagonal crosses? How do solutions change as the faces grow? Shown in the figure is a computer-graphics rendition of a fragment of the

Omega Prism. Renditions of the actual 230 mm × 231 mm × 232 mm prism contain facets so small that they are impossible to distinguish when printed on a page.

The purpose of the figure is to emphasize the difficulty that individuals have when they attempt to count tiles intersected by a diagonal line without actually using a straight edge and drawing the line. When the colors blink, it is impossible for humans to count "intersected" tiles by eye alone.

Difficulty rating: 

Answer: **Ans48**

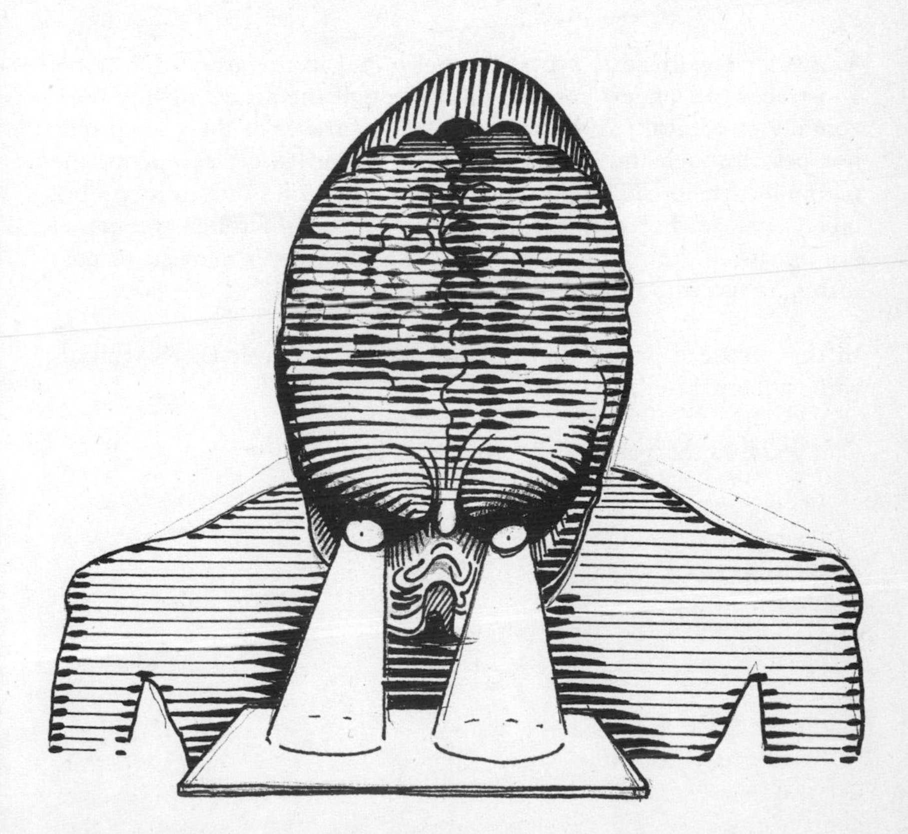

16 ALIEN WORM

> Part of the exam seems to commonly consist of pressing upon a prone abductee's vertebrae one by one from the top of the spine to the coccyx.
>
> C. D. B. BRYAN,
> *Close Encounters of the Fourth Kind*

> The axis of the earth sticks out visibly through the center of each and every town or city.
>
> OLIVER WENDELL HOLMES

Consider a worm that starts at any symbol in the array of symbols and traces the longest possible path through the array, moving horizontally or vertically (not diagonally). As it traverses the grid, it may not pass through the same symbol more than twice, except for the menorah, which can be passed through four times due to its probabilistic nature. For example, if the worm passes through the comet symbol twice during its journey, the worm is never allowed to pass through this symbol again.

In this puzzle, I have found a worm path that traverses 18 symbols while adhering to these rules.

Can you find this path or any other long paths?

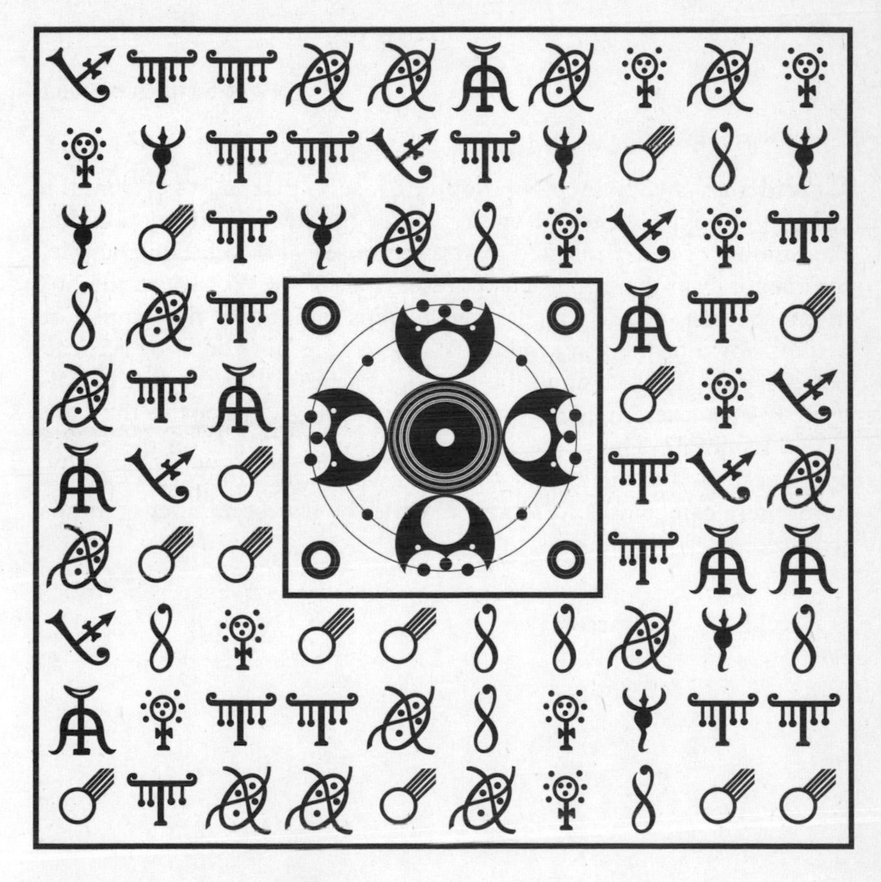

Difficulty rating: 

Answer: **Ans19**

17 ALIEN HOMOPTERA

Em meus sonhos, o peixinho nada.

ARGUS MELLO

$$\frac{\delta^2\psi}{\delta x^2} = -\frac{8\pi^2 m}{h^2}(E-V)\psi$$

ERWIN SCHRÖEDINGER

Consider an alien homoptera (hopping insect) that must get from the cell at the upper left to the empty cell at the lower right by jumping the number of cells indicated by the symbol it lands on. For example, the menorah in the upper left represents a "5" or "6," so on the homoptera's first move, it might move 5 steps right to the fish symbol or 5 steps down to the fish symbol. All moves are horizontal or vertical, not diagonal. Let's assume the homoptera moved to the fish in column 1. From here it could move, for example, two steps to the right to the menorah. The process is repeated until it lands on the empty space at the lower right. Can you find a way to the empty space so the homoptera can gain its freedom? What is the fewest number of jumps required?

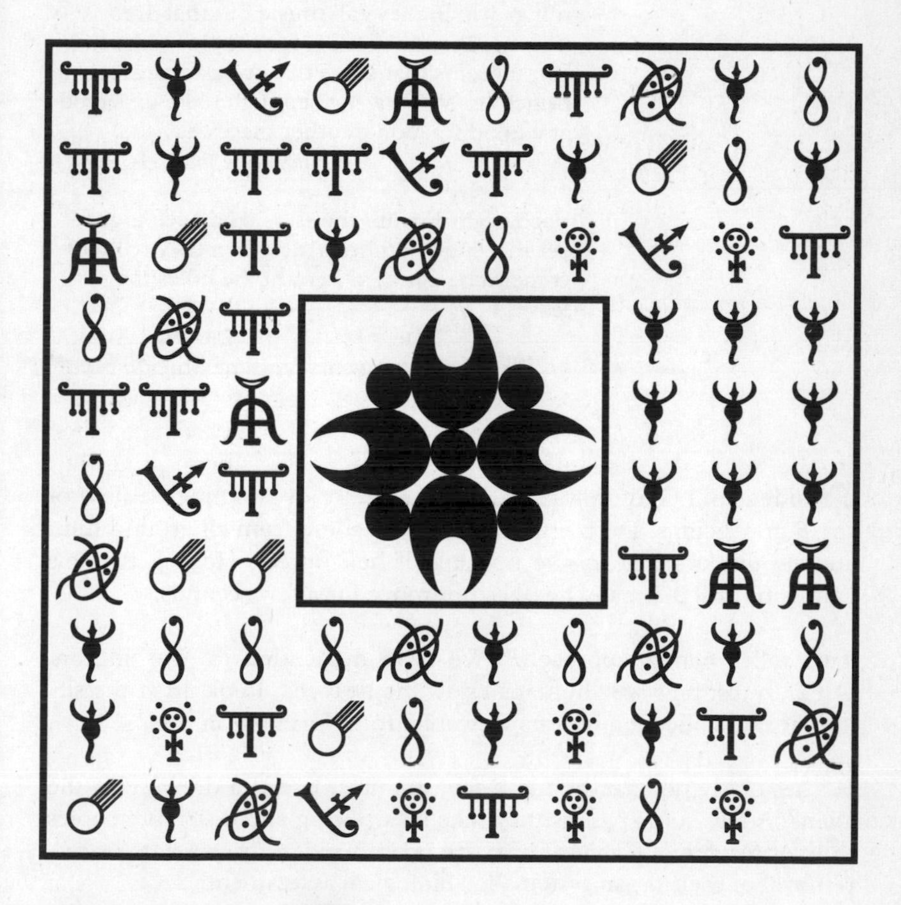

Difficulty rating: 

Answer: **Ans18**

18 STAR CHART

Some other civilizations in space may have no music at all, for various biological or mental reasons. Some others may have vastly different things which they call music but that are incomprehensible to us, for simple reasons. But it seems that some of our basic musical principles are universal enough to be expected at a good fraction of other places.

SEBASTIAN VON HOERNER

Each night I count the stars, and each night I get the same number. And when they will not come to be counted I count the holes they leave.

LEROI JONES AND IMANU BARAKA,
"Preface to a Twenty Volume Suicide Note"

President Bill Clinton wakes up in the center of an immense field of small gray beings. Deep grumbling noises come from all around him, and he thinks he hears the buzzing of helicopters. Hoping that his confusion will dissipate, he does not move for a few seconds.

Two taller beings approach. "We have descended to Washington, D.C., to test how well humans know the heavens. Look up at the sky and draw a collection of stars viewable from Washington, D.C."

He hesitantly nods and, after a few minutes, hands a drawing to the aliens. A few hours pass as they take his drawing and distribute copies to the president's Cabinet, who are asked to identify the stars or risk removal of their organ systems for biological assessment.

Can you identify any of these stars using names that humans have given them? The name of the constellation at right is mentioned in Homer's *Odyssey* of the eighth century B.C. Can you identify the star marked with a question mark? A transmission from this star would take 36 light years to arrive on Earth.

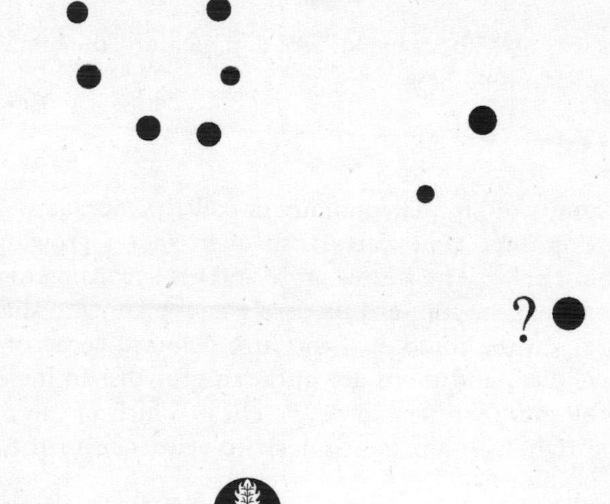

Difficulty rating: 

Answer: Ans47

19 ALIEN SPORES 1

Our civilization is awaiting a telephone call with a difference—placed by people whom we do not know, whom we will not recognize, whom we may not even understand. Would the dangers of answering a call from another civilization outweigh the likely benefits?

IAN RIDPATH, *Worlds Beyond*

Without experience and memory, the brain is mindless.

LEWIS P. LIPSITT

We see two columns of circular containers called xenoplates. The left column contains three xenoplates with alien spores growing in some of the interior circles. The spores grow and die according to certain rules. Starting from the top left, there is a xenoplate with sixteen germinating spores. In the middle of the first column, some of the spore patches have died, and there are only ten growths. In the bottom of the first column, there are fewer patches. Which of the three xenoplates in the right column continues the sequence in the left column?

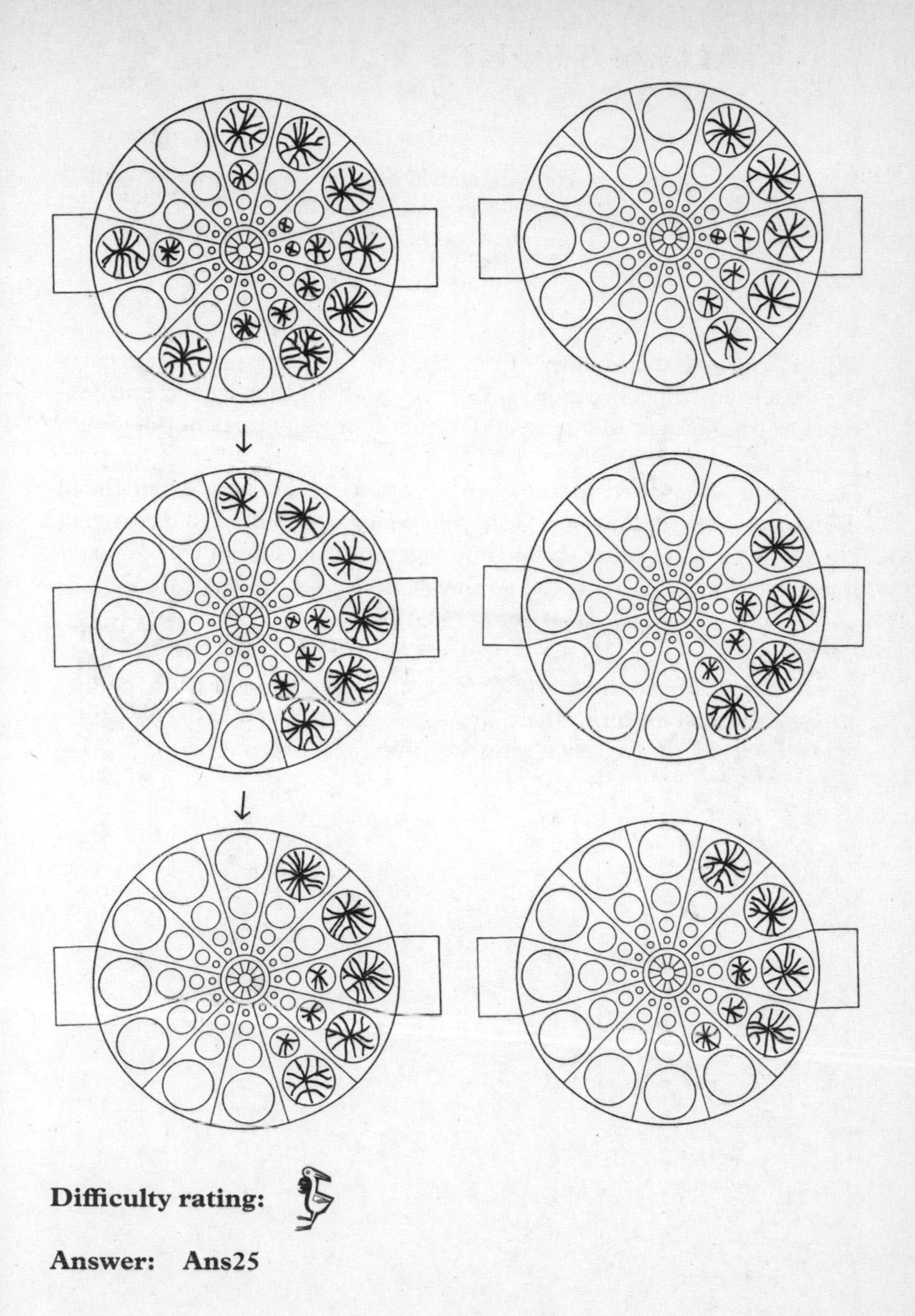

Difficulty rating:

Answer: Ans25

20 ALIEN SPORES 2

Your vision will become clear only when you can look into your own heart. Who looks outside dreams; who looks inside, awakens.
CARL JUNG

Alien spores have escaped from the lab and contaminated rings within a spaceship's hyperdrive. All spores are of the same species except for one. Which ring is coated with a spore that does not belong?

Destroying these alien spores is an easy matter. But each is beneficial to humans, except for one that is genetically engineered to destroy all life on Earth. This killer spore is now camouflaged. Can you discover the alien spore with a mission so that it alone can be eradicated?

Difficulty rating: 

Answer: Ans22

21 ALIEN SPORES 3

> You do not need to leave your room. Remain sitting at your table and listen. Do not even listen, simply wait. Do not even wait, be quite still and solitary. The world will freely offer itself to you to be unmasked, it has no choice, it will roll in ecstasy at your feet.
>
> FRANZ KAFKA

Alien spores have escaped from the lab and contaminated an alien computer's memory arrays. All spores are of the same species except for one. Which spore does not belong?

Difficulty rating: 

Answer: **Ans28**

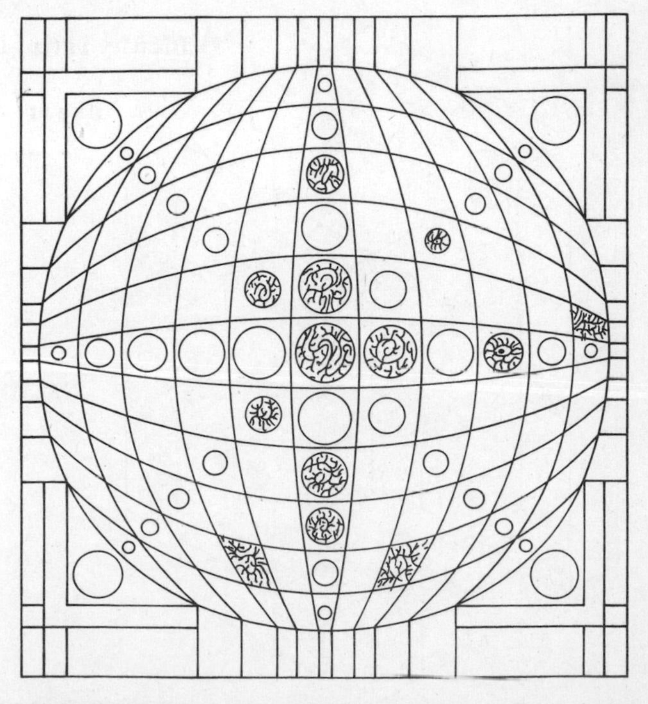

22 ALIEN SPORES 4

> I am half inclined to think we are all ghosts. . . . They are not actually alive in us; but there they are dormant, all the same, and we can never be rid of them. Whenever I take up a newspaper and read it, I fancy I see ghosts creeping between the lines. There must be ghosts all over the world. They must be as countless as grains of sand it seems to me. And we are so miserably afraid of the light, all of us.
>
> HENRICK IBSEN, *Ghosts*

We see an alien food unit. Alien spores have escaped from the lab and contaminated compartments of this unit used to store different kinds of edible plants and animals. All spores are of the same species except for one. Which spore does not belong?

Difficulty rating:

Answer: Ans35

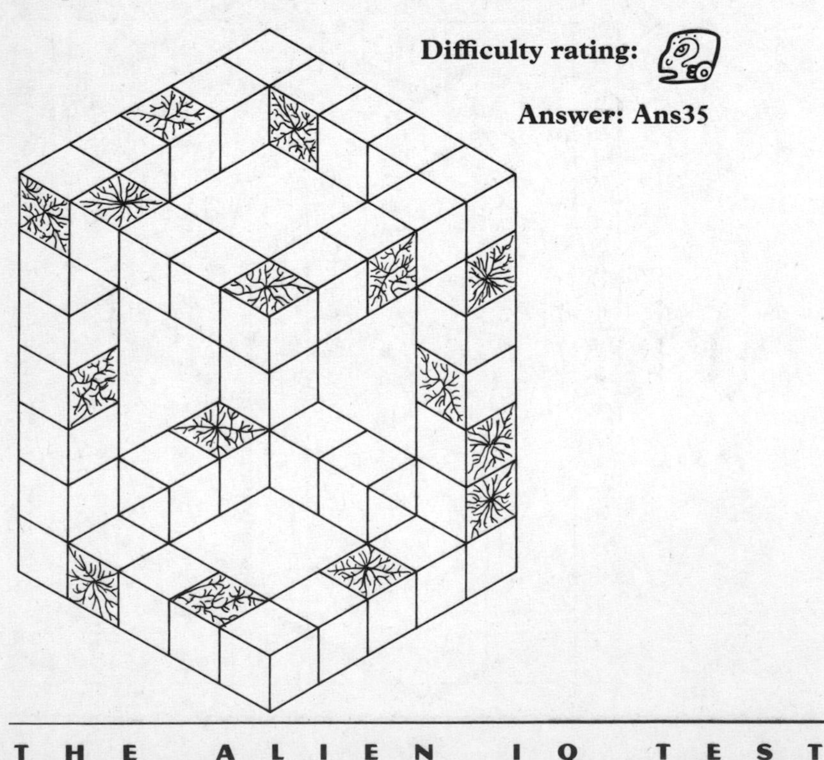

23 ALIEN SPORES 5

O, be prepared, my soul! To read the inconceivable, to scan the million forms of God those stars unroll. When, in our turn, we show to them a Man.

ALICE MEYNELL, circa 1900

Alien spores have escaped from the lab and contaminated an alien hive consisting of three cubes with nine square faces visible in this figure. All spores are of the same species except for those on one of the faces. Which face does not belong?

Difficulty rating: 

Answer: **Ans36**

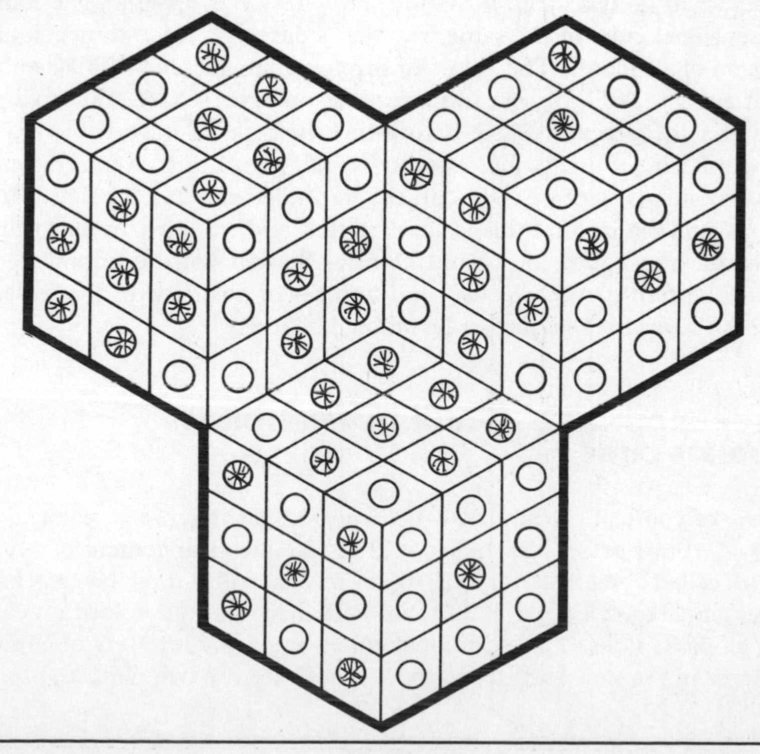

24 RUBIK'S TESSERACT

For Catherine, time had lost its circadian rhythm; she had fallen into a tesseract of time, and day and night blended into one.

SIDNEY SHELDON, *Other Side of Midnight*

The hypercube or tesseract is described by moving the generating cube in the direction in which the fourth dimension extends.

T. BROWNE, *Mystery of Space*

A TESSERACT IN GRAND CENTRAL STATION

Aliens have descended to Earth and placed a 3' × 3' × 3' × 3' Rubik's tesseract in Grand Central Station, New York. (A tesseract is a four-dimensional cube in the same way that a cube is a three-dimensional version of a square.) The colors of this four-dimensional Rubik's cube shift every second for several minutes as onlookers stare and scream. (At first the people believe it to be a new type of advertising from Calvin Klein.) Finally, the tesseract is still, permitting us to scramble it by twisting any of its eight cubical "faces," as described below. Humans soon realize that this is an alien test, and we have a year to unscramble the figure, or Grand Central Station will be annihilated. Your question is: What is the total number of positions of the tesseract? Is the number greater or less than a billion?

BACKGROUND TO A FOUR-DIMENSIONAL RUBIK'S CUBE

Many of you will be familiar with Ernö Rubik's ingenious, colorful 3 × 3 × 3 cube puzzle. Each face is a 3 × 3 cube arrangement of small cubes called cubies. If you were to cut this cube into three layers, each layer would look like a 3 × 3 square with the same four colors appearing along its sides. Two additional colors are in the interiors of all the squares in the first and third layers. These are the colors on the bot-

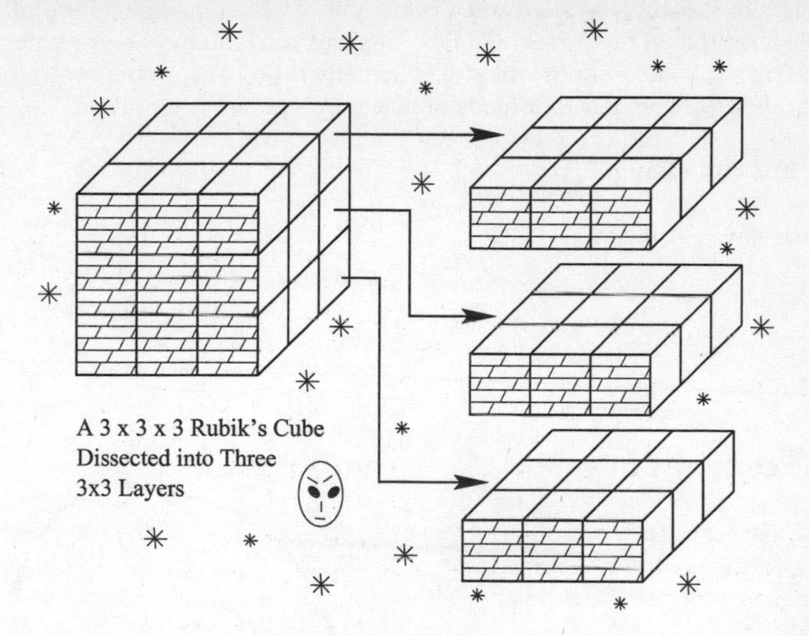
A 3 x 3 x 3 Rubik's Cube Dissected into Three 3x3 Layers

tom of all the squares on the first layer and top of all the squares in the third layer.

The aliens have extended this puzzle to the fourth dimension in which the four-dimensional $3 \times 3 \times 3 \times 3$ Rubik's cube, or tesseract, resembles three $3 \times 3 \times 3$ cubes that are stacked "up" in the fourth dimension. All three cubes have the same six colors assigned to their faces; in addition there are two more colors assigned to the interiors of all of the cubies in the first and third cube. (I refer to the 81 small cubes in this representation as cubies, as have other researchers like Dan Velleman, although each is really one of the 81 small tesseracts that make up the alien Rubik's tesseract.)

The three- and four-dimensional puzzles differ in the following ways. The original Rubik's cube has six square "faces." The Rubik's tesseract has eight cubical faces. In the standard Rubik's cube, there are three kinds of cubies: edge cubies with two colors, corner cubies with three colors, and cubies in the center of faces that have only one color. I ignore the cubie in the center of the cube, which has no color and plays no role in the puzzle. (In fact, those of you who have taken your Rubik's cube apart know that there actually is no cubie in the center.) Rubik's tesseract has four kinds of pieces.

Difficulty rating: 

Answer: **Ans40**

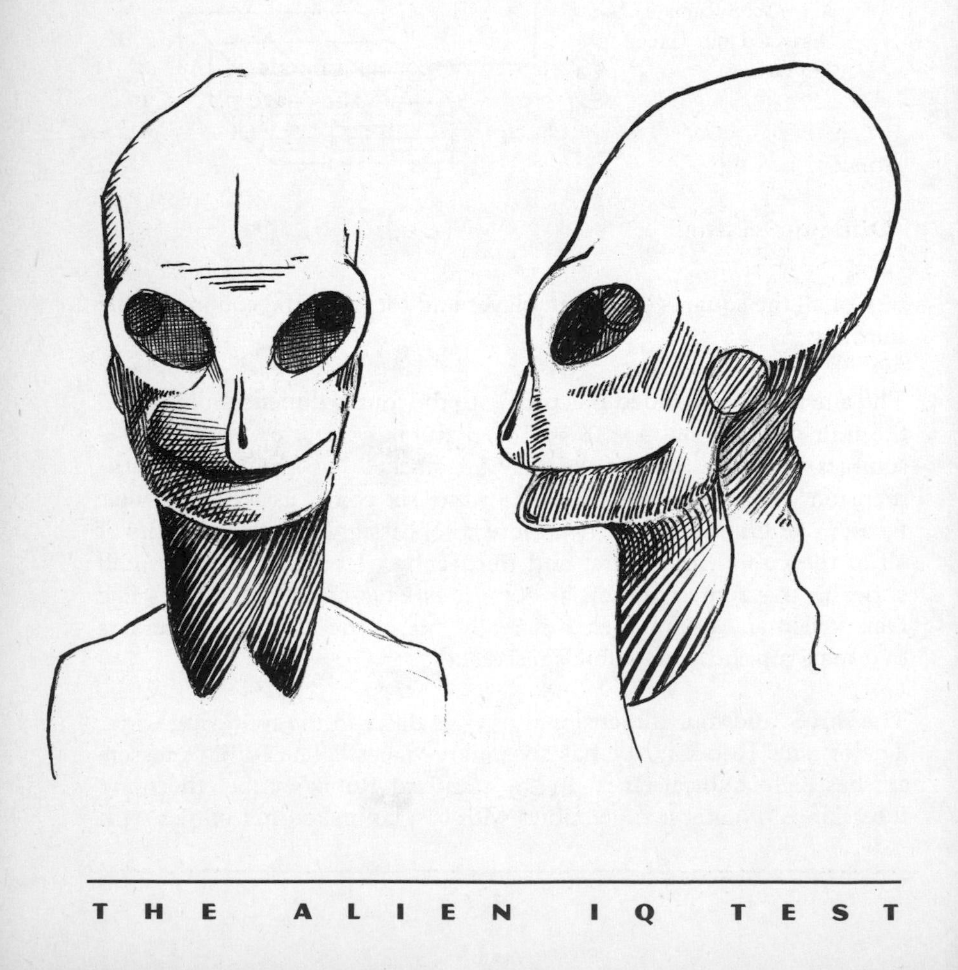

25 ANIMAL EYE

It has been said that when we first discover other civilizations in space we will be the dumbest of them all. This is true, but more than that, we will probably be the only mortal civilization. For it is immortals we will most likely discover. . . .

FRANK DRAKE

In a perfect universe, infinities turn back on themselves.

GERGE ZEBROWSKI, *OMNI*

Aliens wish to assess how well we know other animals with whom we share Earth. To make the task more difficult, they have placed an eye in an alien kaleidoscope to reflect it many times. To what animal does this eye belong?

Difficulty rating:

Answer: Ans39

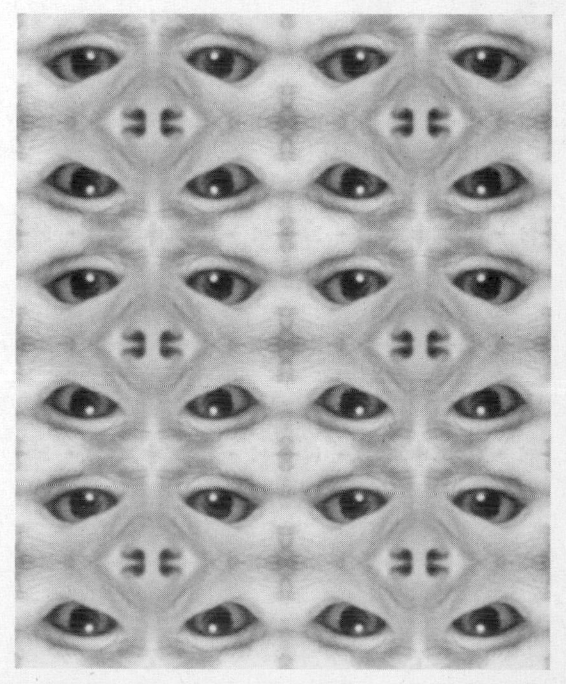

26 COSMIC ROSETTA STONE

> Does it not seem strange to say that His power, immensity, beauty, and eternity are displayed with lavish generosity through unimaginable reaches of space and time, but that the knowledge and love, which alone give meaning to all this splendor, are confined to this tiny globe where self-conscious life began to flourish a few millennia ago?
>
> FATHER L. C. MCHUGH

> Messages in Lingua Cosmica might unceasingly travel through the universe.
>
> HANS FREUDENTHAL, *LINCOS, Design of a Language for Cosmic Intercourse*

Some astronomers and mathematicians believe that symbolic logic would be the vehicle of choice for communications between intelligent beings from different star systems. In the early 1900s, there were several efforts to describe mathematics in a purely logical language. In the 1960s, Hans Freudenthal, professor of mathematics at the University of Utrecht (the Netherlands), attempted to develop a logical language that we could use to communicate with intelligent aliens with whom we have nothing in common.

The language is called Lincos, which stands for "Lingua Cosmica," and it consists of mathematical, biological, and linguistic symbols including some of those employed by earlier mathematical logicians, such as Alfred North Whitehead and Bertrand Russell. Can you decode any of the following message written in the language of Lincos?

Ha Inq *Ha*:

x ∈ Hom.→: Ini.*x*Ext·−:Ini·Cor*x*.Ext˙ =
  Cca.Sec 11×10^{10111}:

∨*x*˙ *x* ∈ Bes.∧: Ini.*x*Ext·−:Ini·Cor*x*.Ext˙>Sec 0:

x ∈ Hom.→˙∨ ⌜*y*.*z*⌝:*y* ⌢ *z* ∈ Hom.∧·*y* = .Mat *x*·∧·*z* =.Pat*x*:

∨*x*˙ *x* ∈ Bes.∧˙∨ ⌜*y*.*z*⌝:*y* ⌢ *z* ∈ Bes.∧·*y* = .Mat *x*·∧·*z* =.Pat*x*:

x ∈ Hom.→˙∧*t*: Ini·Cor*x*.Ext:Ant:*t*.Ant:Ini.*x* Ext˙
  →:*t* Cor*x*.Par·*t* Cor.Mat *x*:

∨*x*:*x* ∈ Bes.∧·∧*t*.Etc:

x ∈ Hom.∧:*s* = Ini·Cor *x*.Ext·
  →:∨ ⌜*u*.*v*⌝*s* Cor*u*.Par·*s* Cor.Mat*x*:
    ∧˙PauAnt.*s*·Cor *v*.Par:PauAnt.*s*·Cor.Pat *x*˙
    ∧:*s* Cor *x*.Uni·*s* Cor*u*.*s* Cor*v*:

∨*x*:*x* ∈ Bes.∧.Etc:

Hom=Hom Fem.∪.Hom Msc:

Hom Fem∩Hom Msc= ⌜ ⌝:

Car:↑*x*·Nnc*x*Ext.∧.*x* ∈ Hom Fem˙
  Pau>˙Car:↑*x*·Nnc*x*Ext.∧.*x* ∈ Hom Msc:

y = Mat*x*.∧.*y* ∈ Hom·→.*y* ∈ Hom Fem:∧:

y = Pat*x*.∧.*y* ∈ Hom·→.*y* ∈ Hom Msc:

x ∈ Hom∪Bes:→:Fin.Cor*x*·Pst.Fin*x*#

Difficulty rating: 

Answer: Ans44

27 ALIEN ANTS IN HYPERSPACE

> [The UFO-abduction cult] threatens the mental health, perhaps even the life, of those unwittingly become participants.
>
> PHILLIP KLASS,
> *UFO Abductions: A Dangerous Game*

Aliens abduct a Kansas City mill worker and, in front of his paralyzed eyes, place a twisting tube within which a robotic ant crawls.

The ant starts at a point marked by a glowing red dot. The aliens turn to the mill worker and say,

> The ant is executing an infinite random walk; that is, it walks forever by moving randomly one step forward or one step back in the tube. Assume that the tube is infinitely long. What is the probability that the random walk will eventually take the ant back to its starting point? You have one week to answer correctly or else we will examine your internal organ systems with a pneumoprobe. However, if you answer correctly, we offer you a reward. We will give Earth the technology to produce rainfall at any time and place desired. As a result, farmers would have bumper crops and food prices would drop.

For days, the mill worker stares at the ant moving back and forth in the thin tube. The only sound is the delicate humming of the ship's engines. From somewhere off in the distance, he begins to smell burning leaves and formalin, and for a moment he dreams of autumn blackbirds occasionally crying amidst vague perpetual clouds. He imagines a world in which one need not move, and birds live forever suspended in space.

You have all the information you need to solve this problem. Can you help the mill worker solve this problem? The ant essentially lives in a one-dimensional universe. How would your answer change for higher dimensions?

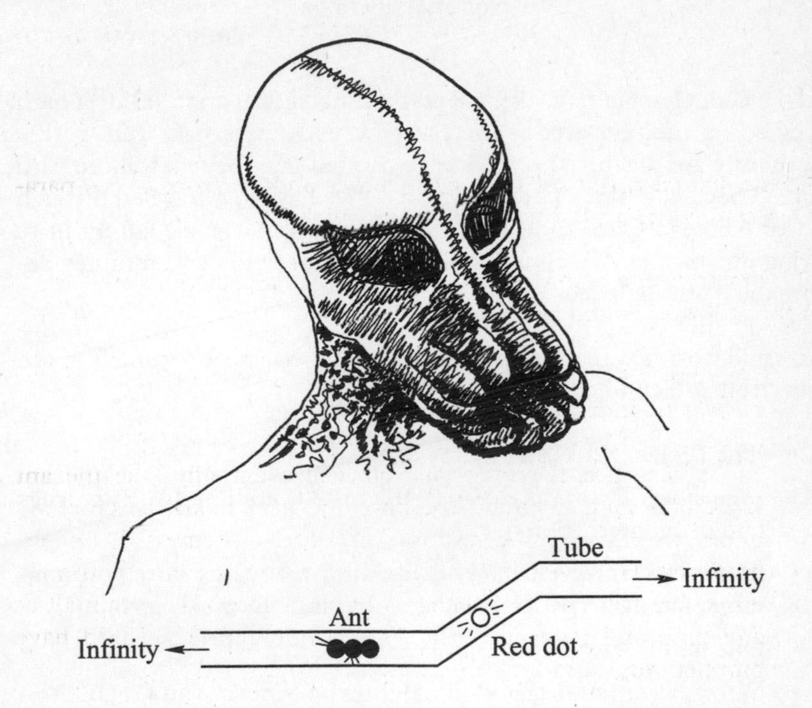

Difficulty rating: 

Answer: Ans50

28 A SEVERED HUMAN FINGER

There is quite definitely something or other deranged in my brain.

VINCENT VAN GOGH

On a cool October day, aliens land their octagonal craft beside a barn located on the outskirts of Kansas City. A Kansas City mill worker hesitantly approaches the craft as two small gray beings walk toward him. They hand him a note that reads, "We have just traveled through a worm hole in space and would like you to select a single gift for us to bring our people. We come in peace, but have only five minutes before the worm hole closes."

The mill worker walks into his barn where he has an assortment of objects from which to choose one potential gift:

- The Bible (Old Testament)
- *Physician's Desk Reference* (PDR), 1990, edition 44 (lists drugs and drug interactions)
- *Mobil 1997 Travel Guide to the North East* (lists hotels, restaurants, family activities, towns, parks, and colorful maps)
- One jar of Peter Pan creamy peanut butter
- *Starry Night* (an original oil painting by Vincent van Gogh)
- Sheet music for Johann Sebastain Bach's *Toccata and Fugue in D-minor*
- ChapStick lip balm
- A Pentium computer
- A severed human finger

When I asked dozens of scientist-colleagues which object they would choose, there was one clear winner. Which object do you think was chosen most frequently? Which do you choose and why?

Difficulty rating: **Answer: Ans51**

29 THE ANTIKYTHERA MECHANISM

> "Am I to be frightened," he said in answer to some report of the haruspices, "because a sheep is without a heart?"
>
> JAMES ANTHONY FROUDE, *Caesar*

On a cool November day, aliens land their octagonal craft beside a Vassar College dormitory located in New York's Hudson Valley. A Vassar College ballet teacher hesitantly approaches the craft, as two small gray beings walk toward her. They hand her a note that reads, "We wish to assess how well your species knows its own history, people, and artifacts. Which of the following terms can you define? We come in peace, but will remove your left kidney for analysis if you do not recognize any of these terms. We do offer you a reward if you recognize a term. We will give you a battery-powered device that stimulates the pleasure center of your brain. There are no electrodes or side effects. Whenever you use the device, you will be more cheerful and optimistic. When set on the highest setting, you will experience intense or paroxysmal emotional excitement."

- The Zeta Reticuli Controversy
- The Art of the Haruspices
- The Antikythera Mechanism
- The Gohonzon of the Soka Gakkai
- The Suckling of the Ainu
- The Matriphagy of the *Diaea ergandros*

The ballet teacher looks at her abdomen and then back at the note, and a tear comes to her eye.

Can you help the ballet teacher by determining which of the above phrases is most recognized by humans? When I asked colleagues and scientists from around the world if they had ever heard of these terms, there was one clear winner.

Difficulty rating:  **Answer:** **Ans53**

30 ALIEN SCRAMBLING

> Know thou that every fixed star hath its own planets, and every planet its own creatures, whose number no man can compute.
>
> BAHA'U'LLAH'

Aliens have coded a message for humanity by scrambling an English sentence. On a cold October day, they descend to Earth in an irregular yellow spacecraft and drop a titanium tablet on the barn of a Kansas City mill worker. The tablet contains scrambled letters. The message reads:

Prowl cove open

The farmer begins to place the letters in different order in an attempt to descramble the message. Day and night he writes words on his barn using chalk from his children. He continues to search for English words using the same letters as in the original message, but so far the resulting combinations have yielded nonsensical sentences, for example:

Nerve coop plow

Vowel prone cop

Colon prove pew

Clone prop wove

Envelop row cop

Propel cone vow

Copper love own

Even after a week, he cannot decode the message, and he fears for his family. Can you help him decode the message? Perhaps the aliens are telling us why they are colonizing our planet. Perhaps they are telling us about their food preferences.

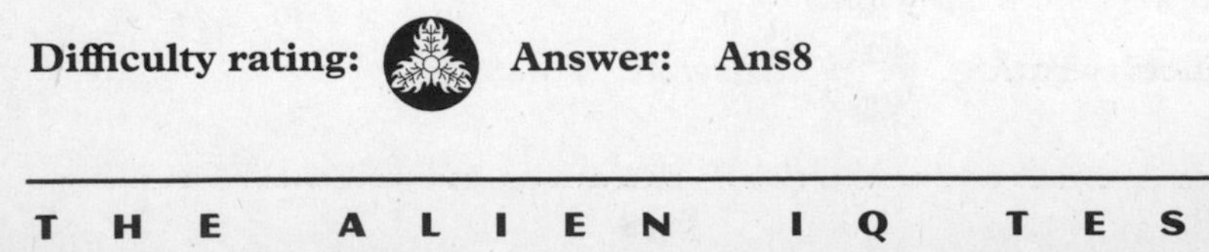

Difficulty rating: **Answer: Ans8**

31 ALIEN AESTHETICS

> For most of this century, scientists have worshipped the hardware of the brain and the software of the mind; the messy powers of the heart were left to the poets.
>
> NANCY GIBBS, *Time*

Aliens wish to understand our concept of beauty. Which one of these items do you think humans rate the most beautiful?

- mossy cavern
- dancing flames
- kaleidoscope image
- snow crystal
- drop of blood
- glimmer of mercury
- scarlet streaks
- mist-covered swamp
- black viper
- retina
- sushi
- uranium
- patriotism
- spiral nautilus shell
- wine
- seagull's cry
- computer chip
- tears on a little girl
- avalanche
- ammonia
- asphalt
- trilobite fossil

The aliens offer you a reward. "If you are able to give the correct answer, we will give Earth our laser scalpels and synthetic skin-graft technology so that facial beauty is affordable to almost everyone. Future operations will be simple, painless, and permanent."

Difficulty rating: 

Answer: Ans55

32 ALIEN KNOWLEDGE AND TALENT

> What matters in the world is not so much what is true as what is entertaining, at least so long as the truth itself is unknowable.
>
> PIERRE LAROUSSE

Aliens descend to Earth and abduct you while your car is stopped at a red light on a dark highway. While aboard their ship, they ask you several questions:

- "Would you rather us give you instantaneous fluency in our language or the instantaneous ability of our greatest musician?" Choose: language or musical ability.
- "We will help you know more about your own race and talents. We will either instantaneously empower you with the musical abilities of the tenth best pianist on Earth or give you instantaneous fluency in ten Earth languages of your choice." Which do you chose, the piano playing or the languages?
- "You may choose ten books from any one field of *human* knowledge, and we will give you perfect memory and understanding of the contents of these books. By 'field of knowledge' we mean books from defined areas such as math, physics, chemistry, geology, business, philosophy, sociology, medicine, law, religion, or other subjects for which colleges might have majors." From what single field of knowledge would you choose the ten books?
- "You may choose ten books from any one field of *our* knowledge, and we will give you perfect memory and understanding of the contents of these books. By 'field of knowledge' we mean books from defined areas such as math, physics, chemistry, geology, business, philosophy, sociology, medicine, law, religion, or other subjects for which colleges might have majors." From what single field of alien knowledge would you choose the ten books?

What are your answers to these questions? What do you think most humans answer? How would your answer to the last question change if the aliens concluded with, "Our warning to you is that no Earthling has ever desired to go home once they acquired this much alien knowledge"?

Difficulty rating: 

Answer: Ans56

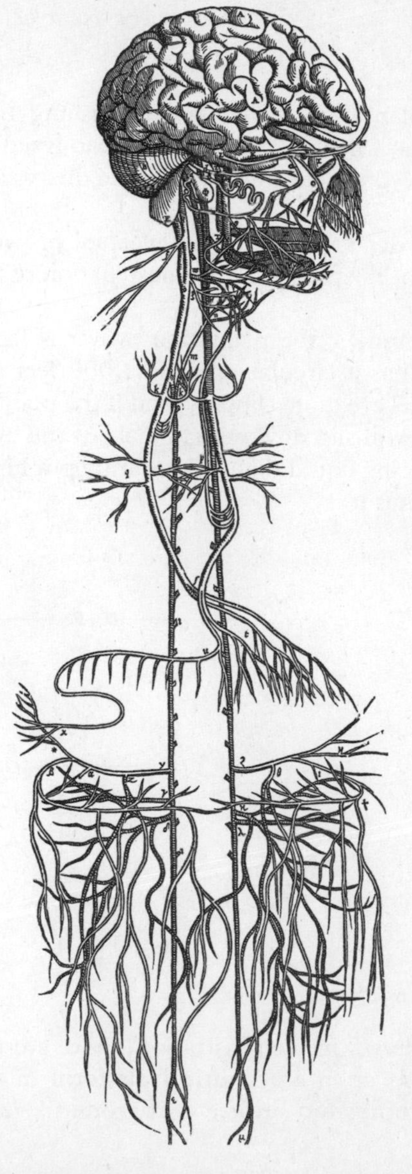

33 THE SAGITTARIUS MANEUVER

Yoboseyo, yi umsik ae daehaeso pyongka lul naeryo bolgayo?

A plantation owner is on a fishing trip north of Belo Horizonte, Brazil, when at 3:30 P.M. he is paralyzed by a flash of light. He is lifted to a cylindrical craft by aliens in dusty silver suits and masks.

In order to assess his intellectual prowess, they transport him to an Earth-like planet in the constellation of Sagittarius.

In minutes, the plantation owner is facing a large in-ground funnel that has a circular opening 1,000 feet in diameter. The walls of the funnel are quite slippery, and if the plantation owner attempts to enter it he will slip down the funnel. At the bottom of the funnel is a sleep-inducing liquid that will instantly put him to sleep for eight hours if he touches it.

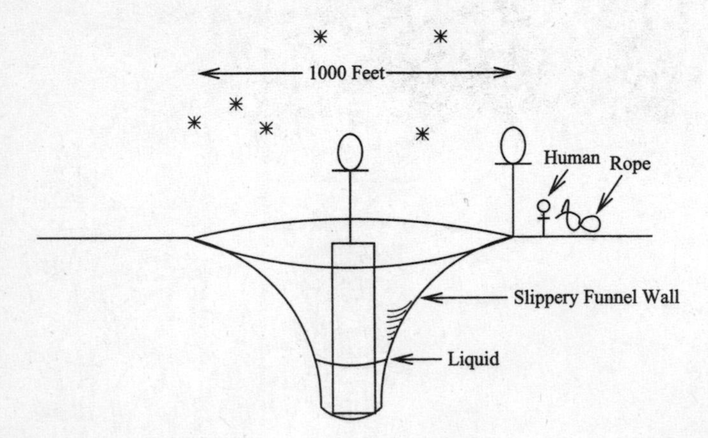

As shown in the illustration, there are two ankh-shaped towers. One stands upon a cylindrical platform in the center of the funnel. The platform's top surface is at ground level. The distance from the platform's top surface to the liquid is 500 feet. The other ankh tower is on land, at the edge of the funnel as illustrated.

The aliens hand the plantation owner two objects: a rope 1,016.28 feet in length and the skull of a chicken. The aliens turn to him and say, "If you are able to get to the central tower and touch it, we will give Earth the cure for cancer. We will also give Earthlings the ability to see in the ultraviolet range, thereby opening up a vast new arena of sensory experience. If you do not get to the tower, we will leave you on this planet after we have implanted a tracking device in your nose. Please note that with each passing hour we will decrease the rope length by a foot."

How can the plantation owner reach the central ankh tower and touch it?

Difficulty rating: 

Answer: **Ans63**

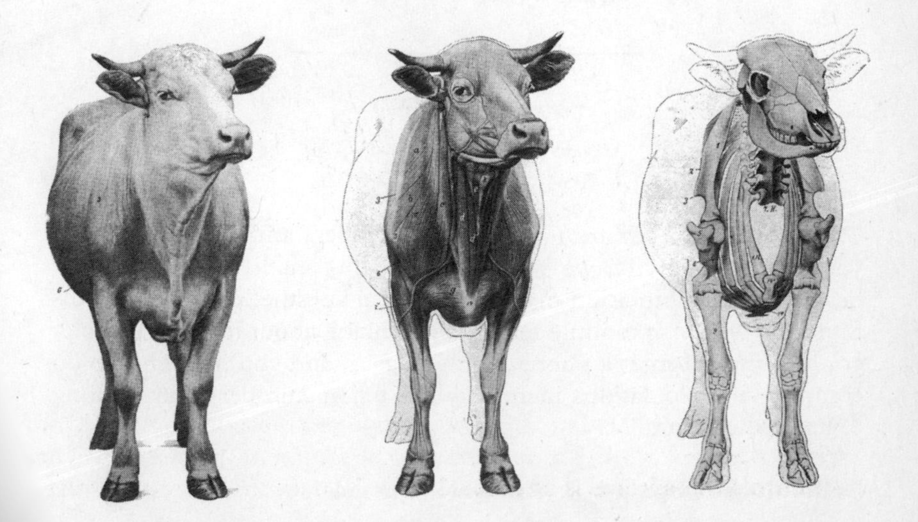

34 SIRIUSIAN GEOMETRY

Madam, vash colchnok vir'val na moi toophel.

Aliens from a planet circling the star Sirius have abducted a Spuyten Duyvil railway engineer and his daughter in order to assess their intelligence. The alien ship is now located at position S in the following diagram. The aliens wish to travel to a wall in space called the Continuum for the purpose of refueling their ship using energy in the Continuum's plasmoid wall.

After this they wish to travel back to a star called Aleph-naught denoted by A. They tell the abductees, "We wish to travel to the edge of the Continuum for refueling then back to Aleph-naught, traveling as short a distance as possible. To what point on the edge of the Continuum should we travel?"

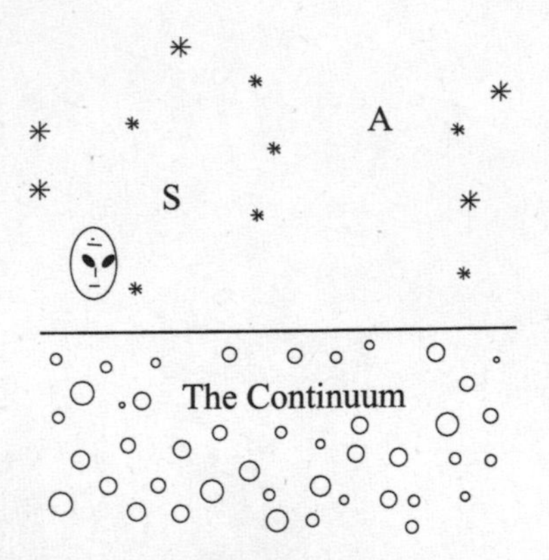

The aliens offer a reward to the railway engineer and his daughter. "If you answer correctly, we will implant a tiny math module in your head, via a small incision made with a local anesthetic. You will communicate with this module merely by thinking about it. With its help, you will never forget a phone number again, and you'll be able to do complicated calculations merely by looking at numbers and wanting to know the result."

Difficulty rating: **Answer: Ans64**

35 HUMAN BRAINS IN A JAR

> The brain is a three pound mass you can hold in your hand that can conceive of a universe a hundred-billion light-years across.
>
> MARIAN DIAMOND

> *Evett mar valaki ebbol az etelbol amit nekem kinalsz vagyen vagyok az elso?*

Sometime between 2:00 A.M. and 3:00 A.M. a restaurant owner is driving a truck filled with pork loins on the A338 trunk road between the villages of Sopley and Avon in Hampshire, England. He approaches a junction near a bridge over the River Avon when his truck lights dim. As he taps on the truck's control panel, he is astonished to see a strange egg-shaped object move from left to right in front of him. An alien emerges from the egg and takes the truck driver to an abandoned railway car that smells of raw meat.

The alien motions to an opaque, formalin-filled jar. "The jar contains one organ, either a kidney or a brain," the alien says.

The restaurant owner watches as the alien drops a brain into the jar, shakes the jar, and quickly withdraws an organ that proves to be a brain. The alien turns to him and says, "What is now the chance of removing a brain?"

Difficulty rating: 

Answer: **Ans65**

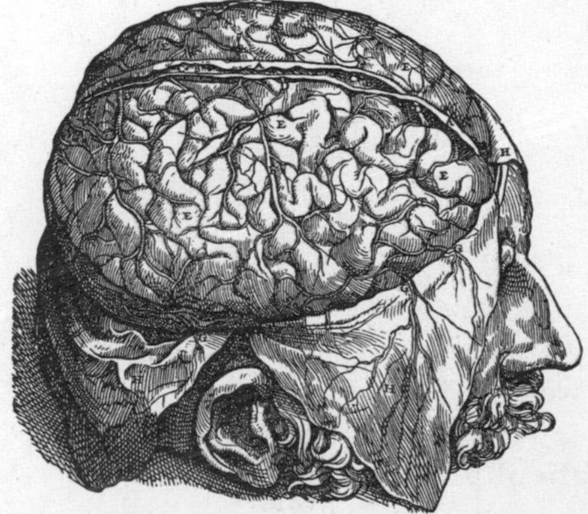

36 HUMAN BELIEF STRUCTURE

> Did we come here to laugh or cry? Are we dying or being born?
>
> CARLOS FUENTES, *Terra Nostra*

> The dreams of men belong to God.
>
> S. R. DONALDSON

Consider two universes. Universe Omega is a universe in which God does not exist, but the inhabitants of the universe believe God exists. Universe Upsilon is a universe in which God does exist, but no inhabitant believes God exists.

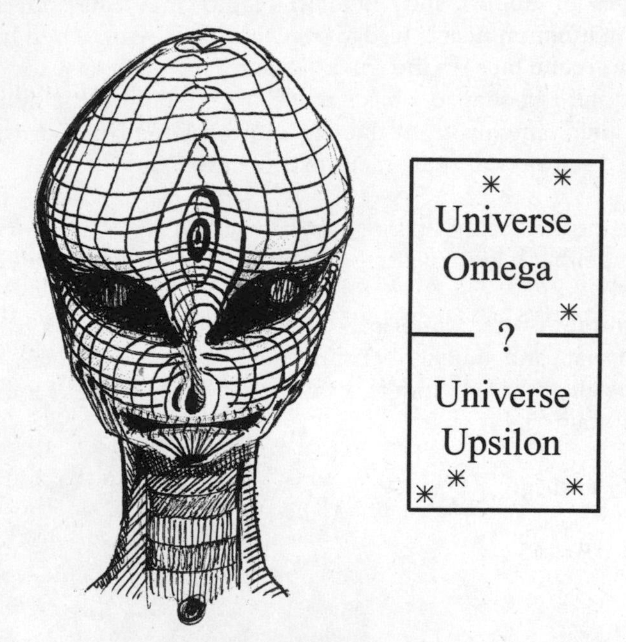

In which universe would you prefer to live?

In which universe do you think most people would prefer to live?

Difficulty rating: **Answer:** **Ans73**

37 CONTACT FROM THE PLEIADES

Often the last thing these abductees want to do is look into the being's eyes.

JOHN MACK, *Abduction*

Rhythm is one of the principal translators between dream and reality.

EDITH SITWELL

At 4:00 A.M. on October 15, fashion model Vanessa Vanderhofen is walking near her home in Hinwel, a canton of Zurich, Switzerland. The air blows the spiral curls of Vanessa's blonde hair out of place just as she sees a silvery, disk-shaped craft in the sky. The UFO swoops down and a figure emerges.

"We are from the Pleiades, a star cluster in the constellation of Taurus, 430 light years from Earth," the alien says. "We wish to assess your intelligence."

The alien gives Vanessa instructions and then places her in a transparent spiral tube one mile in length. The diameter is so small that she must crawl through the tube. She starts at the center of the spiral at 5 A.M. and crawls until she reaches the outlet of the spiral at 5 P.M. She travels at varying speeds, and every now and then pauses, rests, and eats from meager food rations strapped to her belt.

When she arrives at the outlet of the spiral, she rests and then begins her journey back into the spiral the next day at 5 A.M. and arrives at the center of the spiral at 5 P.M.

"You have completed your mission," the alien says. "Now answer a question. What are the chances that there is a location along the spiral that you passed at exactly the same time both days? If you answer correctly, we will explain to you whether we are real or only a hallucination resulting from abnormal electrical activity in your brain's temporal lobes."

Difficulty rating: **Answer:** **Ans74**

38 THE ELK HUNTER'S ABDUCTION

That I should love a bright particular star . . .
WILLIAM SHAKESPEARE,
All's Well that Ends Well

On October 25, 1974, twenty-five-year-old Christie Meier leaves Riverton, Wyoming, in her pick-up truck to hunt elk in a remote corner of Bow National Park. In the forest, she sees seven elk, raises her rifle, and fires at the nearest animal.

Christie then hears something approaching: a tall, gray-skinned being with a large head. Its oval eyes are shiny and expressionless.

"Are you shooting at an enemy?" the alien asks. "Are you at war?"

"No," Christie says.

"Are you intelligent?"

Christie shrugs uncertainly.

"Let me ask you a question." The alien pauses. "Creatures from seven stars are at war with one another. No two distances between any pair of stars are the same. At exactly A.D. 2010, creatures from each star system launch a devastating nuclear arsenal at their nearest neighbor, and they wait to see what happens. None of the star systems has a defense against such awesome fury. Will all seven races be annihilated, or will at least one star system escape injury?"

Difficulty rating: 

Answer: **Ans75**

39 LOSS OF SCIENTIFIC KNOWLEDGE

Science is not about control. It is about cultivating a perpetual condition of wonder in the face of something that forever grows one step richer and subtler than our latest theory about it. It is about reverence, not mastery.

RICHARD POWERS,
The Gold Bug Variations

I cannot persuade myself that a beneficent and omnipotent God would have designedly created the *Ichneumonidae* with the express intention of their feeding within the living bodies of caterpillars.

CHARLES DARWIN

If all human scientific knowledge were lost in a cataclysm, what single statement would preserve the most information for the next generation of creatures?

Difficulty rating:

Answer: Ans76

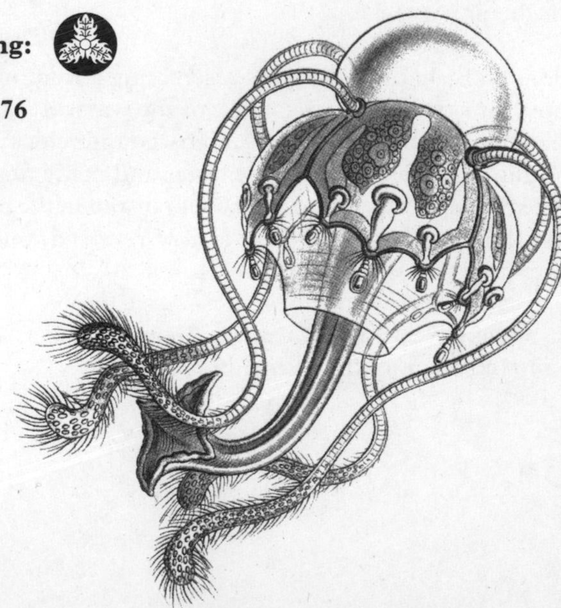

ALIENS AND SPRINKLERS

> Might not tales of alien abduction be attributable to universal human motives? Perhaps some people embrace "alien encounters" for the same reason that others turn to drugs, or alcohol, fetishistic sex, dangerous sports or charismatic religions.
>
> ROBERT SHEAFFER, *Scientific American*

> As they die, the ones we love, we lose our witnesses, our watchers, those who know and understand the tiny little meaningless patterns, those words drawn in water with a stick. And there is nothing left but the endless flow.
>
> ANNE RICE, *The Witching Hour*

A being from the Black Hole Travel Agency, the galaxy's nastiest megacorporation, asks you to solve a problem before permitting you to tour the universe.

Consider an S-shaped apparatus resembling a rotating lawn sprinkler. The apparatus is spun by the recoil of the water it sprays forth. In the top diagram here, the apparatus spins counterclockwise. What happens if this device is placed under water and made to suck water in instead of spewing it out (bottom)? Would it spin in the reverse direction, because the direction of the flow is now reversed, pulling rather than pushing?

If you are unable to solve this problem, the alien will convert the Earth into an alien tourist attraction.

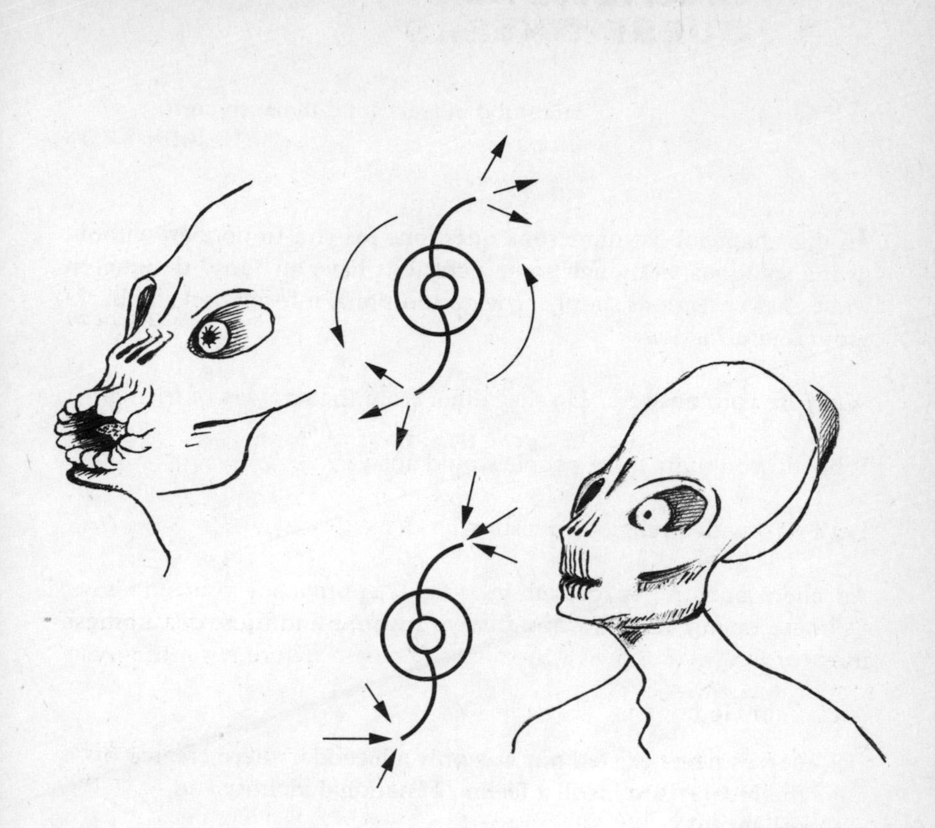

Difficulty rating: 

Answer: **Ans77**

41 UNANSWERED QUESTIONS

She stood in tears amid the alien corn.
JOHN KEATS

In this chapter I list numerous questions for you to ponder, without giving solutions. Although some questions have an "ans" designator, these answer sections simply provide additional information and background information.

What are your answers? Do they differ from the answers of friends?

What do you think most people would answer?

Let's start with a religious question.

An alien lands its spacecraft in a field, approaches you, and says, "Choose one of the following that you would find most disturbing if true."

1. I am God.
2. Moses never existed but was only a legendary hero created by the Hebrews to instill a feeling of national identity and solidarity.
3. Abraham (the father of Judaism and the man who was willing to kill his son Isaac because God told him to) never existed but was merely an invention of the Bible's authors.
4. God exists, but he is a shrimp-like creature living beneath the soil of the planet Mercury.

See **Ans78** for additional commentary.

Another religious question:

An alien lands its spacecraft in a field, approaches you, and says, "In Universe One, there is 'life after death,' but there is no God. In Universe Two, there is no 'life after death,' but God exists. In which universe would you prefer to live?"

The next questions deal with anatomy.

An alien descends to Earth and removes the thumb on every human being. What is the effect on society? How would this effect be different if the thumb were removed from only females, or only males?

Aliens may wish to ask us about the effects of various scenarios on our world. Which of the following would have the least profound effect?

1. A child genius stuns judges at a seventh-grade science fair when he presents a working time machine made from parts of a microwave oven.
2. Scientists discover a beneficial virus that can turn ordinary rocks into protein-rich food. Experts predict the find will lead to the end of world hunger.
3. A meteor the size of a Buick strikes a used-car dealership in Las Vegas. No one is injured in the crash, but the crater opens up a vast underground reservoir of drinking water, solving the desert town's water shortage.
4. A volcanic eruption creates a new land mass that ties the United States to Cuba.
5. Frog legs become the rage in fast-food restaurants.
6. Scientists discover rapidly mutating bees, uncovering evidence that the insects are developing an intelligence that may one day rival that of humans.

See **Ans79** for additional commentary.

Imagine what we might answer if an alien descends to Earth and asks humans the question, "What is an electron?" How should we answer? How could we answer?

Aliens may be entirely different from us. Imagine that aliens from the Zeta Reticuli star system cannot feel pain because they have no pain receptors as part of their nervous system. The aliens descend to Earth and ask humans to answer the question, "What is pain?"

How should we answer? How could we answer?

How would humanity be affected by the presence of an alien with an IQ ten times greater than ours? What new areas of thought might be open to this hyper-IQ individual? What profound concepts or areas of awareness might be available to which we are now totally closed?

A dog cannot understand Fourier transforms or gravitational wave theory. Human forebrains are a few ounces bigger than a dog's, and we can ask many more questions than a dog. Are there facets of the universe we can never know? Are there questions we can't ask?

An alien might ask us a strange question in probability. I call this "The Problem of the Bones," and my mathematical colleagues do not seem to agree on an answer.

Imagine that a creature from outer space walks you before a pit. In the pit are 10,000 leg bones. The creature tells you, "I have cracked each bone at random into two pieces by throwing them against a rock. What's the average ratio of the length of the long piece to the length of the short piece? If you cannot find the solution within two days, I will add your leg bone to the pit."

An alien might ask us a strange question in logic. What would be your response to the following? An alien comes to you and says, "Answer this question with yes or no: Will your next word be no?"

42 MORAL AND EMOTIONAL CHOICES OF HUMANS

> "Tonight," the President of the United States said, "as we gaze at the sky and wonder at what we see there, our minds are crowded with questions."
>
> ROBERT WILSON, *The Harvest*

> Science is the only genuine consciousness-enhancing drug.
>
> ARTHUR C. CLARKE

Aliens descend to Earth and pose several questions.

How would you answer these? You must be truthful.

Do you think most people would agree with your answers?

ORGAN HARVESTING

Researchers at the University of Texas create 125 headless humans by knocking out a gene in developing embryos. With no nostrils or mouth through which to breath, the humans die upon birth. However, their organs may be harvested to save the lives of other infants needing transplants. Would you support the creation of such headless donor stocks? If your infant child needed a transplant, would you allow the life-saving transplant?

AZTECS

In 1487, the Aztec Empire systematically killed 80,000 individuals by extracting their hearts with obsidian knives. These prisoners of war had been caged, fattened, and sedated with a plant called toloatzin so they would not struggle while waiting to be killed. If you could temporarily go back in time to save these individuals, with no risk to your own life, would you do so? Is preventing past suffering less important to you than preventing present or future suffering? Would you be willing to sacrifice a limb to save the 80,000 individuals?

CONODONTS

Conodonts are tiny eel-like creatures that lived 500 million years ago and were our earliest vertebrate ancestors. You have the choice of making the remarkable discovery of a conodont in your neighborhood pond and donating the creature to a local university for study, or winning a sports car in a *TV Guide* sweepstakes. Which do you choose?

DENGUE FEVER

Dengue fever causes horrific joint paint, and it reaches epidemic proportions in Latin America and Caribbean countries. It has already killed 5,000 people. Someone gives you $1,000,000 that you can keep or donate to a physician in Latin America who will, as a result, find a cure for the disease. Do you keep the money or donate it?

COW TONGUES

Molecular biologists at the Magainin Research Institute discover a microbe-killing peptide on cow tongues. This small protein accounts for the fact that tongues rarely get infected and heal rapidly after injuries. If everyday you eat uncooked tongue from a freshly killed cow, you will be guaranteed immunity from all major diseases. Would you eat a tongue every day, knowing that such a practice might wreak havoc on the economy and ecosystem, and be condemned by animal rights advocates and nations with few cows?

NERVE GAS

A Harvard chemist riding on the Tokoyo subway releases a nerve gas chemically related to sarin, the gas used by Iraq to kill its own people, the Kurds. Dozens stagger coughing and choking from the attack in the subway. It turns out that this sarin-like gas does not kill people but rather raises their IQ by 100 points. The Harvard chemist, a member of a religious cult, says that he knew the gas would have this effect.

and that he thought his actions would benefit the world. In fact, the people exposed to the gas make miraculous strides in science, geopolitical diplomacy, and the arts. How should the chemist be punished? Should he be punished?

COLD IMMORTALITY

Aliens release psychrophilic (cold-loving) bacteria that grow best at temperatures close to freezing. The bacteria are engineered to repair cellular damage in humans. In fact, people continually exposed to these bacteria will live forever, but these people must live close to the North or South poles for most of their lives to achieve this benefit. Would you choose to move to colder climates? How would this affect world geopolitics?

MARRIAGE

An alien comes to you and asks whether you would marry it and give up any future romantic contact with humans. You are single, and the alien is kind, intelligent, and loving. The good part about such a relationship is that by touching the alien, you will experience incredible, sensual rapture, and explore marvelous new realms of thought. The bad part is that the alien is an ambulatory sponge accustomed to living in a methane-rich tide pool. If you marry, it must be kept in an aquarium.

Do you marry it? Would your answer change if the alien required no aquarium but rather lived as a symbiont within a pouch in your abdomen?

DREAMING

Marilyn vos Savant is listed in the *Guinness Book of World Records* as having the highest IQ in the world—an awe-inspiring 228. She was once asked, "How can you tell if you are dreaming?" Her response was, "If you're wondering if you're dreaming, you're dreaming."

An alien approaches you and asks you, "How do you know I am real and not a hallucination?" How would you respond?

43 CODED TRANSMISSION

> The ultimate mystery facing us is how matter becomes conscious. Simply put, if we argue that we are made of matter, then how does that matter seemingly produce or create images and thoughts? Or even put more crudely, how does meat dream?
>
> FRED ALAN WOLF

> To a frog with its simple eye, the world is a dim array of greys and blacks. Are we like frogs in our limited sensorium, apprehending just part of the universe we inhabit? Are we as a species now awakening to the reality of multidimensional worlds in which matter undergoes subtle reorganizations in some sort of hyperspace?
>
> MICHAEL MURPHY,
> *The Future of the Body*

Can you decode or determine the significance of this message from the stars?

Teflon harbinger ebon amphioxus lumbricoid indigo ebon nephron Strieber amphioxus rotifer ebon amphioxus moon opal nephron glass UFO Strieber. Teflon harbinger ebon yellow amphioxus rotifer ebon teflon harbinger ebon amphioxus moon permian harbinger indigo opal xenophobic UFO Strieber baboon UFO rotifer rotifer opal wrong indigo nephron glass indigo nephron opal UFO rotifer Strieber amphioxus nephron diamond. Teflon harbinger ebon yellow amphioxus rotifer ebon UFO Strieber. Teflon harbinger ebon yellow vestigial indigo Strieber indigo teflon UFO Strieber indigo nephron opal UFO rotifer diamond rotifer ebon amphioxus moon Strieber. Teflon harbinger ebon amphioxus lumbricoid indigo ebon nephron Strieber amphioxus rotifer ebon amphioxus moon opal nephron glass UFO Strieber. Teflon harbinger ebon yellow amphioxus rotifer ebon teflon harbinger ebon amphioxus moon permian harbinger indigo opal xenophobic UFO Strieber baboon UFO rotifer rotifer opal wrong indigo nephron glass indigo nephron opal UFO rotifer Strieber amphioxus nephron diamond. Teflon harbinger ebon yellow amphioxus rotifer ebon UFO Strieber. Teflon harbinger ebon yellow vestigial indigo Strieber indigo teflon UFO Strieber indigo nephron opal UFO rotifer diamond rotifer ebon amphioxus moon Strieber.

Difficulty rating:

Answer: **Ans4**

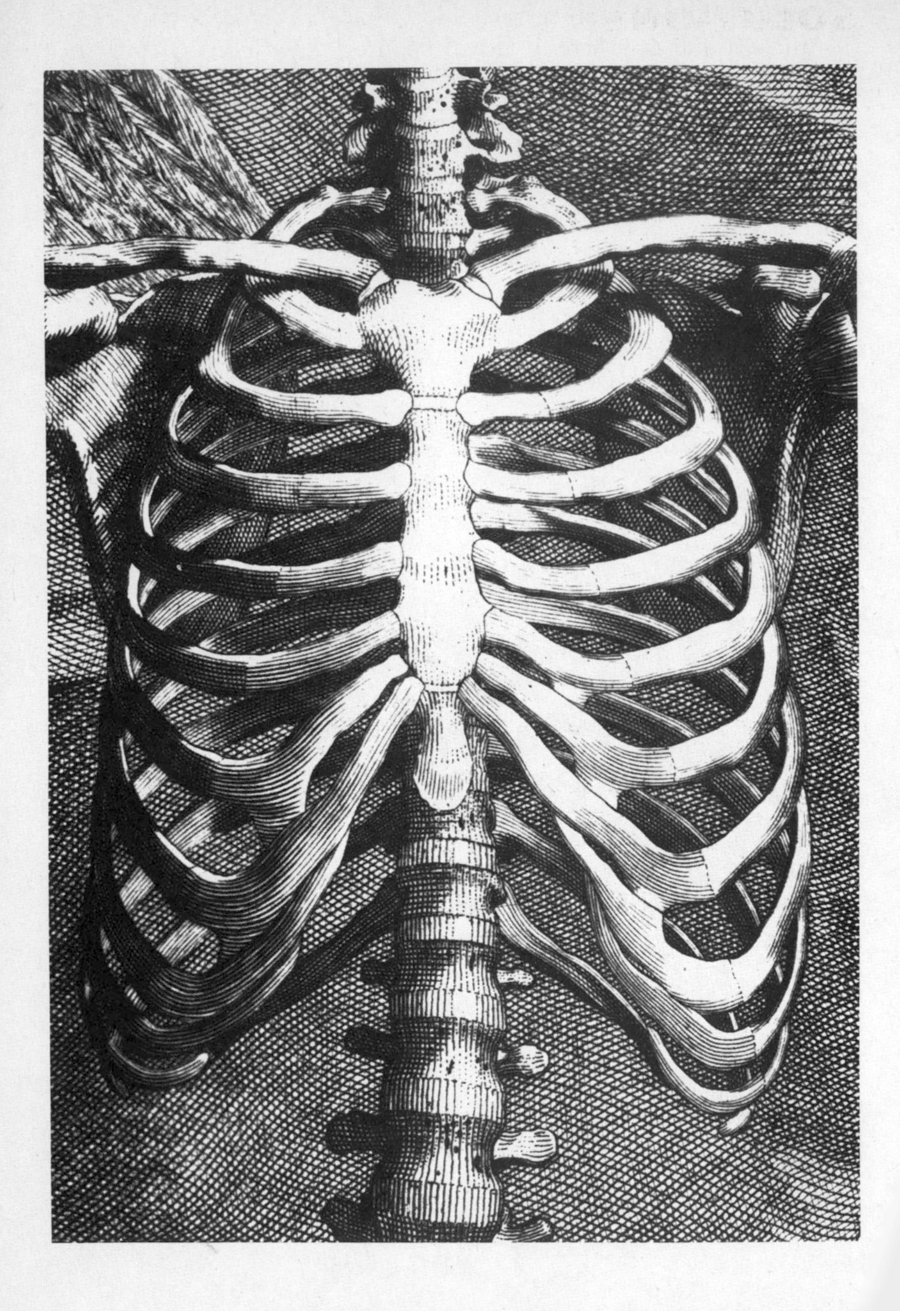

SOLUTIONS

There are times when silence is a poem.
JOHN FOWLES, *The Magus*

ANS1

The missing symbol is the oval.

ANS2

Here is a path that sums to 114.

Is it the shortest path?

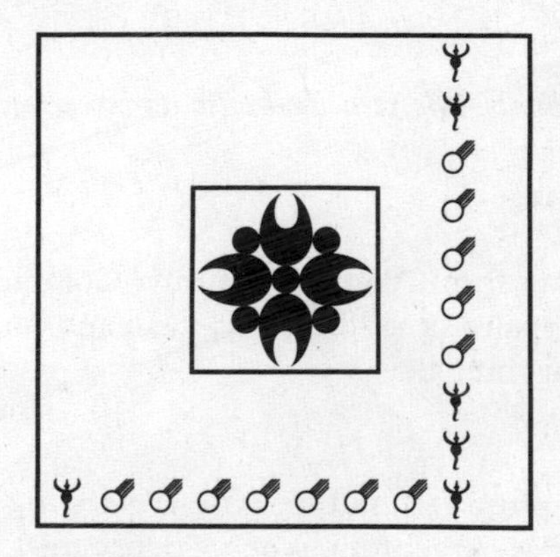

ANS3

Nelson Rockefeller: January 26, 1979, Anne Rice, Isaac Asimov, Richard Marcinko

ANS4

You can decode the message by looking at the first letter in each word.

ANS5

Shah of Iran: Sunday December 2, 1979

ANS6

Area 51, Martin Gardner

ANS7
Nevada Highway 375, T. Phoenix

ANS8
We love popcorn.

ANS9
A chicken farmer in Serra De Almos, Spain.

ANS10
The second tile is the pair that completes the sequence because this tile completes every possible pair of the four symbols. The second tile completes the set of every possible pair of different symbols.

ANS11
Quelle offre raisonable! Je serai heureuse de payer autant, et meme plus si vous voulez!

ANS12
A forestry worker from Alamosa County, Colorado, finds Princess, a two-year-old Appaloosa saddle pony, dead and mutilated after she has been missing for three days.

ANS13
At 7:00 P.M., a physician and his wife are driving to dinner in Maipu near Buenos Aires, Argentina, when a dense mist envelopes their car.

ANS14
Early in the morning, a policeman is at the junction of Highways 3 and 63 near Ashland, Nebraska, when he sees a football-shaped UFO resting on the roof of a house.

ANS15
Here is the largest block of repeated symbols.

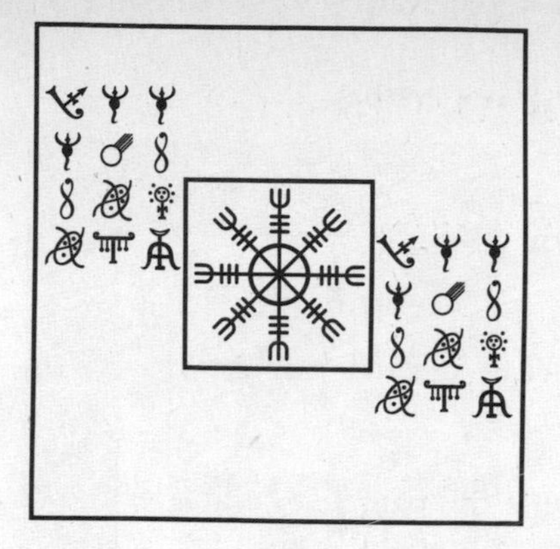

ANS16

The right-most set of symbols is the answer because the ellipses have four points of intersection.

ANS17

The Mayans, who lived in the southern Yucatan Peninsula around A.D. 500, were an unusually gifted and cultured people.

zilfmw 900 z.w., dsvm gsv nzbz ivzxsvw gsvri kvzp, gsvri xrerorazgrlm hfwwvmob xloozkhvw.

ANS18

Below are the cells upon which the homoptera may land to traverse the array. There are probably many other solutions.

How many solutions are there?

ANS19

Below is an 18-symbol path.

ANS20

3, 8, 21, 55, 144, 377, 987, 2584, 6765, 17711, 46368, 121393.

ANS21

Tsv xsrowivm dsln I dzh gzprmt xziv lu uli gsv Csirhgnzh slorwzb dlpv fk lm Csirhgnzh nlimrmt zg zylfg 7:00 zmw yvtzm gl izxv wldmhgzrih gl gsv Csirhgnzh givv rm nb orermt illn.

ANS22

Each ring with spores is linked to one other ring, except for the spore-infested ring near the center which is linked to two.

ANS23

G12124 1675913121213, 132 131135 98 D1594 M67359. I 418 212913 913 A67687, 789972 59787 25198 1712. I 45137 712 1 3121212575 913 P513138212511391 1134 1311012954 913 11828938.

ANS24

GGATCCCAGCCTTTCCCCAGCCCGTAGCCCCGGGAC-CTCCGCGGTGGGCG.

ANS25

The answer is the container in the middle right. A growth dies if it has fewer than two neighbors.

ANS26

The solution is the lower right medallion. Rule: If an even number of circles are blackened, then the center is black. If an odd number of circles are blackened, then the center is not black. The lower right medallion adheres to these rules.

ANS27

21144518 2085 21123151425, 85 919 192091212 22118149147. 2085 92225 418511319 1144 12511419 9142015 8919 19125516.

ANS28

The spores in the center circle do not belong. All other spores can "get" to at least one other spore by moving like a Knight in chess.

ANS29

The answer is the "man." On each row, the first figure plus the second figure minus the third figure is the fourth figure.

ANS30

414213562373095048801688724209698078569671875 37.

ANS31

2089986280348253421170679821480865132823066470938446095 5058223172.

ANS32

A Neanderthal jaw, from Kebra cave in Israel, clearly shows the characteristic lack of a chin and the space behind the last molar.

Tsrh hkzxv dzh kozxvw rm gsv qzdh lu Nvzmwvigszoh yb zorvm erhrglih.

Fli 50,000 bvzih, Nvzmwvigszoh orevw hrwv-yb-hrwv drgs nlwvim sfnzmh rm z hnzoo ozmw.

ANS33

Drawn below is one possible solution. Can you find any others?

ANS34

Dear Secretary Forrestal: As per our recent conversation on this matter, you are hereby authorized to proceed with all due speed and caution upon your undertaking. Hereafter this matter shall be referred to only as Operation Majestic Twelve.

ANS35

The spores on the 4th block from the bottom in the left support do not belong. Every other spore patch can get to at least one other spore patch by moving along 3 adjacent faces. (Moves are orthogonal.)

ANS36

The face with four contaminated circles does not belong. All other faces have a prime number of contaminated circles. (A prime number cannot be written as the product of two smaller factors greater than 1. 5 and 3 are primes but 4 and 10 are not because $4 = 2 \times 2$ and $10 = 5 \times 2$.)

ANS37

It continues to be my feeling that any future considerations relative to the ultimate disposition of this matter should rest solely with the Office of the President following appropriate discussions with yourself, Dr. Bush, and the Director of Central Intelligence.

ANS38
Harry Truman, Dana Scully, Fox Mulder, John Loengard

ANS39
Monkey

ANS40
The total number of positions of Rubik's tesseract is 1.76×10^{120}, far greater than a billion! The total number of positions of Rubik's cube is 4.33×10^{19}. If either the cube or the tesseract had been changing positions every second since the beginning of the Universe, they would still be turning today and would not have exhibited every possible configuration.

ANS41
http://sprott.physics.wisc.edu/pickover/home.htm

ANS42
This shows the entire portal (liver) vein freed from all parts to which it is joined (gall bladder, liver, stomach, spleen, omentum, mesentery, and intestines).

ANS43
http://sunsite.unc.edu/lunar/alien.html

ANS44
Hans Freudenthal's language is explained in his book *Lincos: Design of a Language for Cosmic Intercourse*, published in the Netherlands in 1960. The three-letter symbols are derived from Latin roots. For example, "Fem" means "female." "Msc" means "male." Here is a decoding of the message: The existence of the human body begins some time earlier than that of the human itself. The same is true for animals. Mat, mother. Pat, father. Before the individual existence of a human, its body is part of the body of its mother. It has originated from a part of the body of its mother and a part of the body of its father.

ANS45
http://www.seti-inst.edu

ANS46

The third set in the second row is the answer because it has three different sizes of ellipses. All of the initial examples in the problem have three different sizes.

ANS47

The premiere star of constellation Bootes (*boo-oh-teez*) is Arcturus, only 36 light years away from Earth. There are several stars closer to Earth than Arcturus such as Tau in the constellation Cetus. It is only 12 light years away and is a solitary yellow star like our sun. Epsilon, in the southern constellation Indus, is also a good candidate for an advanced civilization. It is 11 light years away and similar to our sun.

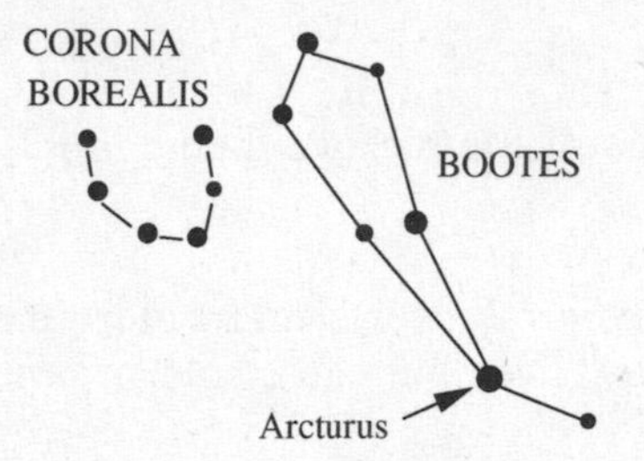

ANS48

The number of tiles a diagonal crosses is the length of one side of a face plus the length of the other minus the greatest common divisor (GCD) of the sides' lengths. The greatest common divisor of two integers is the largest number that divides both integers. For example, a 231×93 face would have $231 + 93 - 3 = 321$ crossed tiles because 3 is the greatest common divisor of 231 and 93. Recall that when a diagonal crosses the exact corner of a tile, it crosses fewer tiles than when it crosses at a point not at the corner.

In the $230 \times 231 \times 232$ prism given, we have three different possible combinations of rectangular sides denoted by A, B.

In particular, we can list values for A, B, GCD, and number of squares cut by the diagonal: (230, 231, 1, 460), (230, 232, 2, 460) and (231, 232, 1, 462). Values of GCD = 1 correspond to prism sides that yield the most cut squares when traversed by a diagonal. The figure shows a plot of those values of A and B that yield GCD = 1, and therefore this plot visually indicates which side lengths should be used to create the most cut squares. To produce this figure, GCD is computed for $1 < A < 200$ and $1 < B < 200$. The density of black dots is fairly uniform, and the wonderful complexity of the plot belies the apparent simplicity of my Omega Prism puzzle.

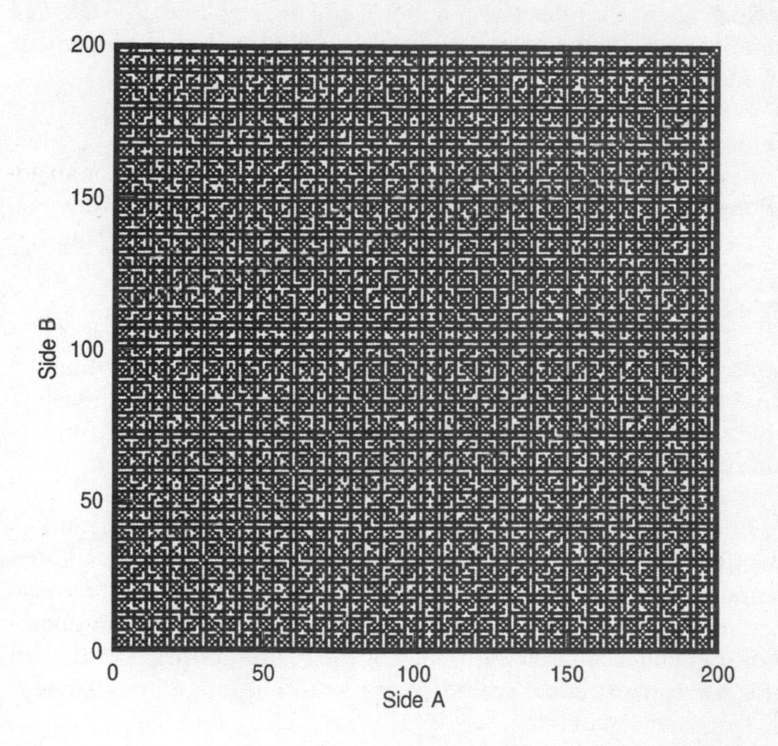

ANS49

About 41% of the respondents chose the bottle of cologne, 23% the book of matches, 17% the rodent, 11% the pen, and 5% the pin.

Reasons given for selecting the cologne

Starting fires and treating wounds and insect bites (because some colognes contain as much as 80% alcohol), cooling the skin, fighting (using shards of glass from broken bottle), bartering (if natives are present on world), and attracting potential humanoid mates. The glass could also be used to reflect or focus light to start fires, and for cutting.

The bottle could be drained and used as a canteen.

The odor might ward off predators.

Some of the cologne components, such as balsams, gums, oleo resins (all which retard the rate of evaporation), and higher aliphatic aldehydes, geraniols, citronellols, and benzaldehydes may function as a pheromone for some of the indigenous wildlife.

Reasons given for selecting the pen

Marking items, leaving a trail, disassembling and using as a straw, using as a blow gun, flattening end to make a knife, and taking notes.

Reasons given for selecting the rodent

Testing potential foods and water sources before eating and drinking them, consumption (if large rodent, e.g. tapir), and companionship. Problems: How would you keep the rodent without a cage? Also, you could only poison him once. Then you're without a food tester.

Bets Lillo commented, "In addition to the rodent offering the ability to test for poison in potentially edible vegetation or contamination in potentially drinkable fluid, it would offer some company. In a world where nothing else was familiar, even a rodent might be some companionship and a tangible reminder of the reality of earth in an alien world. A pregnant rodent would be the best."

Reasons given for selecting the Rubik's cube

Diversion and intellectual stimulation. If aliens were secretly observing you, choosing this intellectual item may impress them sufficiently to release you early.

Dency Hutchings comments, "We may be able to devise a code so that a cube manipulation such as (R+, U-, D+, B-) means something like 'lead me to water.' More sophisticated manipulations, such as solving the cube from any scrambled state, would likely mean something more profound such as 'cook me a nice thick steak, medium rare please.'"

Reasons given for selecting the matches

Starting fires. Problem: Would such a small number of matches really be useful over the course of an entire year? Martie Saxenmeyer would use the fire for many purposes: ". . . to make a hardened pointed stick with which to dig, to spear animals or fish, to strip leaves to weave a shelter or a basket for cooking or storing food, and to dig for shell fish."

She would also use the fire for: "cooking food; keeping warm; cleaning animal skins for clothing and shoes; scaring away night predators; cleaning or sterilizing scrapes, scratches, or water; hardening the bones of animals for making implements of survival; providing charcoal with which to write on flat stones or leaf-like vegetation (to maintain sanity)."

Reasons given for selecting the remote control

The remote's small internal parts may be useful; the battery could be used to start a fire using a very fine wire. If aliens were secretly observing your key presses, you might be able to convince them of your intellectual worth by typing various mathematical identities, prime numbers, etc.

Reasons given for taking the pin

Bending the pin and using as a fish hook, using as a surgical instrument, and removing splinters.

ANS50

Mathematical theoreticians tell us that the answer is one—infinite likelihood of return for a one-dimensional random walk. If the ant were placed at the origin of a two-space universe (a plane), and then executed an infinite random walk by taking a random step North, South, East, or West, the probability that the random walk will eventually take the ant back to the origin is also one—infinite likelihood. Our three-dimensional world is special: Three-dimensional space is the first Euclidean space in which it is possible for the ant to get hopelessly lost. The ant, executing an infinite random walk in a three-space universe, will eventually come back to the origin with a .34 or 34% probability. In higher dimensions, the chances of returning are even slimmer, about $1/(2n)$ for large dimensions n. The $1/(2n)$ probability is the same as the probability that the ant would return to its starting point on its second step. If the ant does not make it home in early attempts, it is probably lost in space forever. Some of you may enjoy writing computer programs that simulate ant walks in confined hypervolumes and making comparisons of the probability of return. By "confined" I mean that the "walls" of the space are reflecting so that when the ant hits them, the ant is, for example, reflected back. Other kinds of confinement are possible. You can read more about higher dimensional walks in: Asimov, D. "There's No Space Like Home." *The Sciences*, 35(5)(Sept/Oct 1995): 20–25.

ANS51

The Vincent van Gogh painting was selected most often as the gift of choice. Many feared giving the aliens advanced technology or any information (such as the finger, which would help them understand our biology) that would help them conquer Earth. Those who feared the aliens the most suggested giving them the peanut butter or lip balm.

None of the respondents chose the Bible as a gift of first choice.

Reasons for choosing the painting

The painting is visual and requires no translation. Elizabeth Poole comments, "Since the note is in English, this means that the aliens have been watching us for a long time, and already know about our culture and our language. They know so much that we'd better give them a real gift. There's only one irreplaceable, truly precious thing on that list: van Gogh's *Starry Night.* I'd hate to so impoverish my own world, but better safe than sorry."

Jim McLean comments, "They ask for a gift, and one of the best paintings ever by one of the greatest artists ever seems an appropriate gift for such a momentous occasion. The subject of the painting also seems to resonate nicely with the event."

Reasons for choosing the Bach musical score

Peter Andrews comments, "I'd choose the Bach score because it's not obviously overly informative (although I have real concern about pa-per and ink analysis), but it's not meaningless. They should certainly be able to conclude we have an understanding of patterns, symmetry, and mathematics. They may even conclude we have a sense of beauty and be appreciative of the gift. You'd have to have a real death wish to give them the Pentium or the finger."

Alrin Anderson comments, "If we give the musical score, I think they'd just use cryptoanalysis to look for a message. At least it would keep them busy."

Reasons for choosing the peanut butter

Bets Lillo comments, "George Washington Carver's development of the peanut as a food source was instrumental, not only in providing a low cost nutritional item to support the South's population, but also in demonstrating the contribution of someone who only years before would have been confined to perform manual labor. Peanut butter would symbolize the hope that we have as Americans for this kind of constructive social evolution—by enlarging our social view to accept more people into society, we enrich all of our lives in ways that we cannot predict nor imagine."

Greg Kishi reminds us that the peanut butter would give the aliens significant information about animal life on Earth because the peanut butter would contain traces of DNA from peanuts and from the oils (corn, soybean, or coconut oil), plus DNA from various insects (400 parts per 16 oz. jar), rats, and humans whose fingerprints on the jar contain DNA.

Mike Hocker comments, "The octagonal ship is probably secretly monitoring the Earth at the same time the alien is asking for the gift from the mill worker. The ship is absorbing all the electromagnetic spectrum (air waves) as it waits. Little robots are probably scurrying around collecting all sorts of interesting things (grass, bugs, dandelions). The outside of the ship behaves like flypaper collecting everything from dust to bacteria."

Reasons for not choosing the Bible

Martha L. Saxenmeyer comments: "I would not choose the Bible, especially the Old Testament, for several reasons. First, while the

Judeo/Christian tradition is a strong one in the United States and in Europe, it is not the primary belief system for most of the world.

"Second, the Old Testament philosophy is primarily one of law and retribution. (Exodus 20 records something like: 'I the Lord Thy God am a jealous God, visiting the iniquity of the fathers upon the children of the third and fourth generations of them that hate me . . .')

"Third, the Old Testament, at least the Pentateuch, is the document of an enslaved and repressed people. While it is a part of history, it is not how I would like an alien culture to view our culture.

"Fourth, the Old Testament defines a class structure, based on birth rather than achievement. This is demonstrated by the 'Priests and Levites' being considered more holy than others. It is demonstrated also by the captivity of the Jews, by the prejudice against them (Joseph and others), and by the assumption of superiority by the Egyptians. Since the Old Testament is a biased viewpoint, and since it does not define us in a proper light, I would not present the Old Testament as my gift."

Reasons for not choosing the severed finger

Martha L. Saxenmeyer comments: "While presenting a severed human finger might be a way of helping the aliens to understand our human physiology, I would not choose this. First, it might cause the aliens to believe that human beings can regenerate. Second, it might cause the aliens to believe that human beings had little regard for their own bodies or lives—that they desired to be dismembered or killed. Third, it might be perceived by the aliens as food. Fourth, it might make the aliens believe that we had little regard for each other. Fifth, it makes me squeamish."

ANS52

Energy spiders, Francis Gary Powers, Soledad O'Brien

ANS53

Eighty percent of those surveyed did not recognize any item on the list.

By far the most commonly recognized term was Ainu. A few learned individuals recognized three items on the list.

- *The Zeta Reticuli Controversy* relates to a star map drawn by Betty Hill, an alien abductee from the early 1960s. She claims to have seen a star map in an alien spaceship. Subsequent analysis of Betty's drawing of the map led some to identify the stars as from a double-star system called Zeta Reticuli, 37 light years from the sun and one of only two known double-star systems thought to be able to sustain life.
- *The Art of the Haruspices* refers to entrail reading—divination by studying the shapes of internal organs, such as the liver of killed animals. Ancient Mesopotamians, Greeks, Romans, Etruscans and many others practiced this art. Etruscans, for example, read messages in the surface folds of the livers, veins, spleens, lungs, and hearts of sheep.
- *The Antikythera Mechanism* is a metal device with gears and gear trains found in a 2000-year-old shipwreck near the coast of Greece. Before the discovery of the intricate mechanism, no comparable artifact from this period had been found. It is currently thought that the device functioned as an astronomical computer that indicated the relationship of certain planets and star constellations.
- *The Gohonzon of the Soka Gakkai* is a ritual drawing (mandala for meditation) used by the Nichiren-sho-shu Buddhists of Japan.
- *The Ainu* are a near-extinct Caucasian race of Japanese. The Ainu once had pale skin, brown hair, short body structure, and were extremely hairy. In 1977, there were only about 12,000 Ainu in existence, most of them indistinguishable from other Japanese due to interbreeding. In early times, female Ainu would suckle bear cubs. The Ainu sometimes exhibited compulsive mimicry as part of their religious fervor. In the past, women often tatooed mustaches around their mouths.
- *The Matriphagy of the Diaea ergandros* refers to Australian baby spiders who eat their mothers as a natural part of their life cycle. After the babies hatch, the mother becomes a "living refrigerator" for the babies who begin to suck on the blood from her unresisting leg joints. After a few weeks, they consume her entirely.

Suggestions by colleagues for terms that should be added to the knowledge test list: Steatopygia, the Popul Vuh, Uracil, Fulcrum, Fovea centralis, Alpha Ursa Minor, *Laissez-faire,* Acetylcholine, Adenosine triphosphate, modified tit-for-tat, the Book of Chilam Balam, Clotho/Lachesis/Atropos, Peltier effect, Seebeck effect, Kerning, Paija, Cheirognomy, Elytra, Glossolalia, Il-Moran, the Tunguska Event, the Cambrian Explosion, the Ediacara Fauna.

Bets Lillo comments on the original list of terms, "Why did the aliens believe that knowledge of these items was so critically important? I'd rather see the aliens come to Earth in search of some sort of 'goodness' instead of administering an IQ or an awareness test. If these aliens are clever enough, I would think they could discern if the ballet teacher is kind to her family and treats her neighbors in a just and fair manner."

Mike Hocker responds to Bets Lillo, "But what is 'goodness'? How does one really define 'just and fair manner'? Humans can't even decide these issues. Humans can largely agree on some broad boundaries. It is not good to eat the neighbors' children, even if doing so gives one's own children a better chance to survive (note that the aliens might think eating the neighbors' children is a perfectly acceptable act). The aliens could do much better than trying to enforce an arbitrary standard of 'goodness' by simply enforcing honesty, i.e., the expungement of deliberate dissemination of untruths and partial truths. A kidney per lie would tend to dissuade most people."

ANS54

The aliens are among us. Oprah Winfrey, January 29.

ANS55

The following is the list of terms sorted in order of beauty as determined by scientists and colleagues I surveyed. The numbers in parentheses indicate the number of times the term was an individual's first choice: dancing flames (31), snow crystal (20), mist-covered swamp (17), spiral nautilus shell (10), mossy cavern (5), kaleidoscope image (5), avalanche (4), computer chip (3), seagull's cry (3), tears on a little girl (3), trilobite fossil (2), glimmer of mercury (2), wine (2), asphalt (1).

Many colleagues suggested additions to the "beauty" list, for example: horses, smiling little girl, Horsehead nebula, rings of Saturn, snow-capped mountain in a forest, palm trees on a Caribbean beach, newborn human infant, looks that pass between couples who have been happily married for 30 years, the feel of a hug and the sound of an "I love you" from a small child, the feeling of coming home after a long journey, a familiar smell (jasmine or honeysuckle on a summer evening breeze, baking bread) that conjures up a rich and sweet memory, human body, beehive, coral, butterfly wings, the grace of an animal moving, music of great composers, bird songs, baby's smile, valley of wildflowers, sunset, Earth from 30,000 feet, members of your preferred gender with big secondary sexual traits, *Mona Lisa,* Eiffel tower, *Guernica, Romeo and Juliet,* "Ode to a Grecian Urn," *The Symphonie Pathetique,* a starry night with no light pollution, full moon close to horizon, deep forest at evening, "my children," ice covered trees after a freezing rain, sunset/sunrise from an airplane, a couple on a couch drinking coffee while looking out at a snowstorm, children laughing, Andrew Lloyd Weber's *Phantom of the Opera* or *Les Misérables,* rainbow, dawn, springtime flowers, autumn trees (not autumn leaves), *Afternoon of a Faun* by Debussy, Jaclyn Smith, the scent of lilacs, sex, the Grand Canyon, Bryce Canyon, a hillside of deciduous trees in autumn, the *Overture of 1812,* waltzes, Dean's gold-medal performance in ice dancing in the Olympics, birds (in their better aspects), a happy baby, freedom, waves crashing against a beach, Milky Way on a dark clear night, crescent moon in twilight, a woman's face, a beautiful Sierra mountain range, a redwood forest in the great Pacific northwest, a beautiful woman (preferably smiling), a cold glass of beer, Beethoven's *Op131 C minor String Quartet,* Vermeer's painting of girl in blue dress with jug, pounding surf, fields of grain, livestock grazing in the pasture, the triangular patterns on a shell called *Cymbiola innexa,* Great Pyramids, Niagara Falls, the human eye, redwood trees, windswept pines on a ridge, clouds seen from above, Claudia Schiffer, mountains in autumn, Grand Tetons behind Jenny Lake, sunrise view with setting near-full moon, long exposure color photo of Andromeda Galaxy, "cow shirr shang hu-AY yo ju-AH ga ma?," the spectrum, emeralds on black velvet, olympic-level figure skating pair, poetry, *Jesu, Joy of Man's Desire* by J. S. Bach, orange/cream cheese chocolate chip cookie, Christie Canyon's breasts, a black night sky sprinkled with star dust, a newborn baby's laugh of joy, young female nude, soap bubble, velvet, rose, snow leopard, the "conversation" of wolves, pictures of maltese dogs or silky terriers in motion, Arabian horses or panthers running, New Jersey or San Francisco ocean waves breaking on a beach, still photographs of some dog or cat breeds in "proud full stance," a hand reached out to help, the formula $e^{i\pi} = -1$.

Peter Andrews comments, "I can answer the question for most interesting, but for beauty I need to understand the presentation: scale, context, lighting, my physical environment. Beauty is in the eye of the beholder. And even a snowflake holds no beauty in the dark."

Bets Lillo comments, "With the exception of ammonia, wine, and a seagull's cry (and possibly sushi) these are primarily items which would be only sensed visually. I think beauty has to command a deeper call than just the eyes—to somehow engage the heart. It is not the thing that is itself beautiful, it is the feeling it calls out within us. . . . "

Brad Pokorny comments, "Both the computer chip and snow crystal are beautiful because of their complexity, order, and diversity—all within a small package. Both say volumes about the beauty and intelligence of their creators."

Herman Harmelink comments, "I chose the tears on a little girl because I feel that tells an entire story. Beauty is in the complexity, not just in the visual aspect, but rather in everything real or imagined that can be attached to it."

ANS56

After conducting extensive surveys, I find that humans greatly prefer fluency in an alien language to alien musical ability. (A ratio of 9:1 respondents preferred language to music.)

I find that humans prefer fluency in ten Earthly languages to extreme piano-playing talent. (A ratio of 3:1 respondents preferred the languages to music.)

Here are the most commonly chosen Earthly knowledge fields, sorted from most frequently given response to least frequently given response: physics, mathematics, philosophy, medicine, psychology, sociology, history, languages, business, literature, religion, electrical engineering, chemistry, and botany.

Here are the most commonly chosen alien knowledge fields, sorted from most frequently given response to least frequently given response: physics, philosophy, mathematics, engineering (including aerospace engineering), sociology, biology, religion, history, and literature.

Dave Glass asks: "How much of a civilization's culture is reflected in the engineering (and architecture) of the culture? Would it be possible to analyzc a civilization based solely on their engineering accomplishments and derive what their thoughts, values, and life styles were?"

Don Webb comments, "I would chose the following ten Earthly languages: Mandarin, Cantonese, Spanish, German, French, Japanese, Russian, Tamil, Navaho, and Arabic."

ANS57

Spicy tekka maki. Katie Couric, Larry King.

ANS58

http://home.earthlink.net/~pleja

ANS59

Roswell, New Mexico

ANS60

What's the frequency, Kenneth?

ANS61

Robert Sheaffer in a 1995 issue of *Scientific American* says, "Aliens seem positively obsessed with human sexuality. Why should the creatures show virtually no interest in studying our cardiovascular, lymphatic, or digestive systems but instead concentrate practically all their attention on our genitals? Such aliens sound very much like inventions of our own minds."

ANS62

To be certain that the alien has two animals of the same species, he must let drop four animals—one more than the number of different species.

To be certain he has a male-female pair of the same species, he must let drop 12 animals—one more than the total number of animal pairs.

ANS63

The plantation owner ties the rope around the base of the ankh tower on the edge of the funnel, and then carries the rope on a walk around the funnel's edge. As he completes half his walk around the circular aperture, the rope begins to wrap around the central ankh on the cylindrical platform, and when he reaches his starting point he ties the other end of the rope to the ankh on the edge of the funnel. Having created a rope bridge from ankh to ankh, he can pull himself across. The chicken skull serves no purpose.

ANS64

Suppose that the aliens aim for some point P on the edge of the Continuum. Reflect the aliens' original position on the edge of the Continuum. Then the distance SPA equals S'PA, and the latter will be a minimum when S'PA is a straight line. It follows that P is the point such that SP and PA make the same angle with the edge of the Continuum.

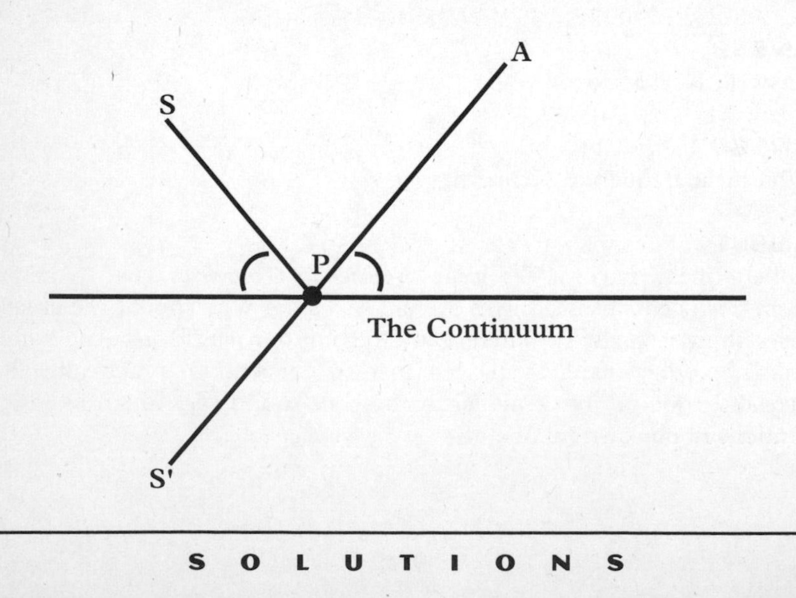

(Note that Heron of Alexandria [A.D. 75] used a similar argument to conclude that when light is reflected from a mirror, the angles of incidence and reflection are equal.)

ANS65

The probability of removing a brain is now quite high, much higher than 50%. The chance is 2/3 or 66.6%. Denote the brain added to the jar as "brain 2." After removing the brain from the jar there are three equally likely states: brain 1 in jar, brain 2 outside of jar; brain 2 in jar, brain 1 outside of jar; kidney in jar, brain 2 outside of jar.

In two out of three cases, a brain remains in the jar. Let me clarify by adding my assumptions: When the alien reaches into a jar containing a brain and a kidney, assume it is equally likely that he will withdraw a brain as he will withdraw a kidney. Also, it is equally likely that the jar initially contains a brain or a kidney.

ANS66

Ah I gfimvw gsv mvcg xlimvi, I hfwwvmob uvog z dzin szmw rm nb ldm. I svhrgzgvw, uli nlgrlm dzh mvziob fmvmwfizyov. I gfimvw zmw ulfmw nbhvou hgzirmt rmgl gsv ilfmw vbvh lu z kivggb dlnzm drgs yildm szri. Ssv ollpvw rmgl nb vbvh, kzfhvw uli zm rmhgzmg, zonlhg hnrovw, zmw hzrw: "I mvvw blf."

ANS67

119447720249892581203851665820676436622934188700177088 360.

ANS68

Bill and Hillary Clinton

ANS70

I believe your esophagus has turned into a functional pizza.

ANS71

Sheryl Crow and Sharon Stone

ANS72

One night after my first encounter, I had a second and final dream of the two gray aliens. A shiver went up my spine as I looked into the aliens' glistening eyes. I felt a chill, an ambiguity, a creeping despair. The small beings were still. None of us moved.

ANS73

I surveyed fifty individuals. Universe Omega and Upsilon were chosen by roughly equal numbers of people. Some respondents suggested that if people think God exists, then God is sufficiently "real." Other individuals suggested that people would behave more humanely to each other in a Universe where people believed in God. (However, couldn't one counter that an ethical system that depends on faith in a watchful or vengeful God is fragile and prone to collapse when doubt begins to undermine faith?)

Here are some comments I received.

D. Reese said, "I choose Upsilon. I've had enough of living in Omega already, thank you very much."

L. Miro said, "I vote for Upsilon; if God exists there are many possibilities to discover this existence. The idea of communing with a god that doesn't exist seems scary."

P. Andrews said, "There have been many 'gods' that I wouldn't want to be in the same room with, so how believers define God is critical. Another important question is: Does God provide for life after death?"

K. Daniels said, "I would prefer to live in Universe Omega. If the God of Upsilon is omniscient, He will know of the existence of Universe Omega and its believers, and perhaps save them."

D. Winarksi chooses Upsilon because, "Many wars have been fought over religion, with each side believing that God was on their side. Maybe life would be treated more preciously if death was regarded as final."

E. Poole said, "I'd far rather live in the universe where people believe in God. I figure they stand a better chance of being better-behaved toward their fellow beings. In the universe where nobody believes in a god that actually exists, everybody goes to hell anyway, right?"

B. Lillo said, "If the inhabitants of Omega interpret the physical signs available to them as evidence of God, then does it really matter if God exists? These people would follow their belief with the semblance of moral order which would be in harmony with the physical phenomena they observe—thus, their religion would be helping to synchronize them with the natural order and to make their lives better off for their belief.

"Concerning Upsilon, if God exists but the inhabitants cannot discern that existence, then one would assume that their moral order would be assembled based on what they observe. If what they observe is a manifestation of God's order in the universe, then it should not matter whether they believe in God or simply observe and adapt to the universe around them.

"The only aspect of material difference between these two scenarios would be the question of afterlife. Again, this premise would be larger than a question of belief. . . if what really happens is re-incarnation, then it will happen whether one believes or not.

"If there is some sort of otherworld afterlife that is a function of the existence of God, then the question becomes one of the method God would use for judging worthiness. If God requires in Upsilon the belief in Him/Her in order to gain entry to the afterlife, then it would be pointless, as no one believes. God would require enough logic and order that He/She would never set up an algorithm in which the answer would be the null set. If God's algorithm were different (some measure of living a life of peace, love, acceptance, and inclusion), then it wouldn't matter whether one believed in God so much as whether one were able to assimilate and practice living according to some qualitative measures that would be visible to God, with or without the individual's belief."

M. Saxenmeyer said, "A belief in God is likely to influence the behavior of the inhabitants.

"Those who believe in a God of the Judeo-Christian tradition—or of most of the Eastern Religion traditions—will most likely pattern their behaviors after what they perceive as 'the good.' If the God were one of love and tolerance, then the pattern of behavior in that universe would likely be of love and tolerance.

"Therefore, it is more likely that the Omega universe would be a more secure and enjoyable place to live."

ANS74

The chances are 100% that there was a spot along the spiral that Vanessa passed at exactly the same time both days. To understand this, visualize Vanessa's trips taking place at once. She starts from the center at the same time her "twin" starts from the spiral outlet. At some point along the journey, they must meet as they pass each other. That will be the place and time.

ANS75

With seven star systems, at least one will always survive. There will be two star systems closer together than any other star pair, and these must have fired their missiles at each other. Continue this line of reasoning and you will see why at least one star system must survive.

What are the maximum and minimum number of survivors for related problems using different numbers of stars? How are your answers affected for two-dimensional, three-dimensional, and four-dimensional arrangements of stars?

ANS76

This question was asked by the great American theoretical physicist Richard Feynman. His answer was, "All things are made of atoms—little particles that move around in perpetual motion, attracting each other when they are a little distance apart, but repelling upon being squeezed into one another."

ANS77

The sprinkler does not turn at all. In the standard sprinkler, the water sprays out in organized jets. However, when water is sucked in, there are no jets. The water is not organized. It enters the nozzles from all directions and therefore applies no force.

(Note that this problem really was studied by nuclear physicists, quantum theorists, and even pure mathematicians both at Princeton University and at the Institute for Advanced Study in the late 1930s. This question was studied with particular zeal by the theoretical physicist Richard Feynman.)

ANS78

According to M. Levin's, F. Maranz's, and R. Ostling's *Time* article "Are the Bible's stories true?" (146[25] Dec. 18, 1995: 65): "Most scholars suspect that Abraham, Isaac, and Jacob, Judaism's traditional founders, never existed; many doubt the tales of slavery in Egypt and the Exodus; and relatively few modern historians believe in Joshua's conquest of Jericho and the rest of the Promised land. There is no [archeological evidence] that Moses existed. Nothing suggests that the Israelites were ever wandering in the desert after fleeing Egypt."

ANS79

All of these ideas, from the 7th grader building a time machine to the mutating bees, came from psychic predictions printed in tabloid newspapers. The predictions were supposed to come true in 1995. Sources: *National Enquirer, National Examiner,* and *Weekly World News.*

DEAR HUMAN: We have interrupted transmission of this book at the printing stage to add "extra" answers, which are themselves hidden enigmas, puzzles, and coded transmissions for you to decode and ponder.

—Message received from an alien communication

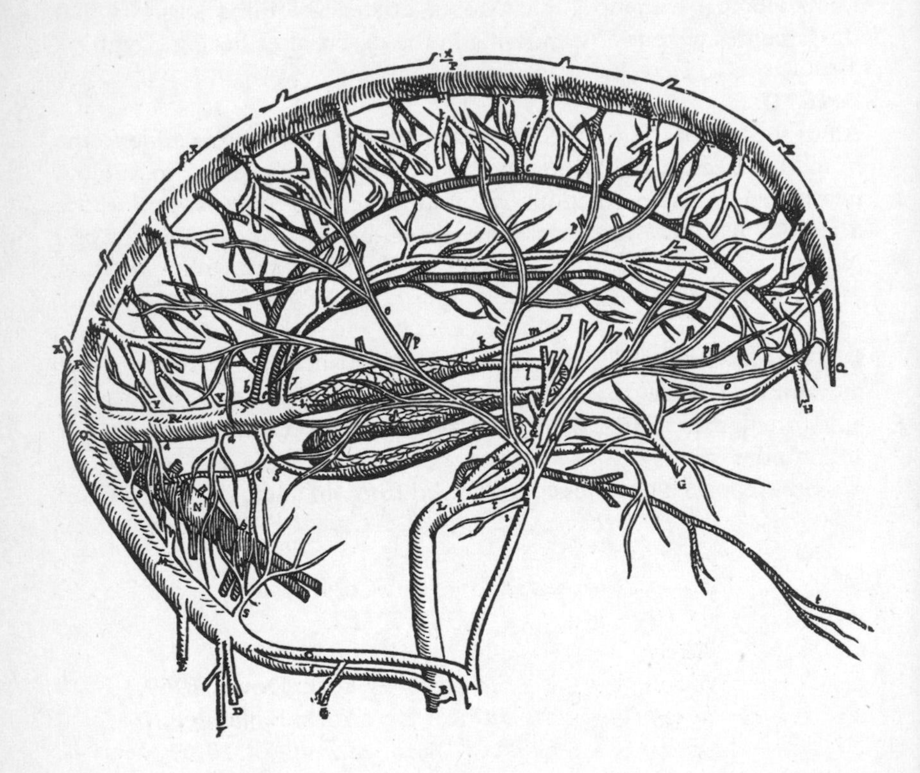

FOR FURTHER READING

Blackmore, S. "Alien Abduction." *New Scientist*, 144(1952) (Nov 19, 1994): 29–31.

Brookesmith, P. *UFO: The Complete Sightings.* New York: Barnes and Noble Books, 1995.

Bryan, C. D. B. *Close Encounters of the Fourth Kind: Alien Abduction, UFOs, and the Conference at M.I.T.* New York: Knopf, 1995.

Emery, C. "Alien Autopsy: Show-and-Tell." *Skeptical Inquirer*, 19(6)(Nov/Dec 1995): 15–16.

Emery, C. "John Mack: Off the Hook at Harvard, but with Something Akin to a Warning." *Skeptical Inquirer*, 19(6)(Nov/Dec 1995): 4–5.

Frazier, K. "UFOs Real? Government Covering Up? Survey Says 50 Percent Think So." *Skeptical Inquirer*, 19(6)(Nov/Dec 1995): 3–4.

Freudenthal, H. *Lincos: Design of a Language for Cosmic Intercourse.* Amsterdam: North-Holland Publishing, 1960.

Gillon, E. *Geometric Design and Ornament.* New York: Dover, 1969.

Hopkins, B. *Intruders.* New York: Ballantine, 1987.

Klass, P. *UFO Abductions: A Dangerous Game.* Buffalo, New York: Prometheus Books, 1994.

Klass, P. "The GAO Roswell Report and Congressman Schiff." *Skeptical Inquirer*, 19(6)(Nov/Dec 1995): 20–22.

LaPlante, E. *Seized.* New York: HarperCollins, 1993.

Lehner, E. *Symbols, Signs and Signets.* New York: Dover, 1969.

Mack, J. *Abduction* (Revised Edition). New York: Ballantine, 1995.

Nickell, J. "Alien Autopsy Hoax." *Skeptical Inquirer*, 19(6)(Nov/Dec 1995): 17–19.

Pickover, C. *Chaos in Wonderland.* New York: St. Martin's Press, 1994.

Pickover, C. *Black Holes: A Traveler's Guide.* New York: Wiley, 1996.

Raymo, C. *365 Starry Nights.* New York: Fireside, 1982.

Sheaffer, R. "Truth Abducted." *Scientific American*, 273(5)(Nov. 1995): 102–103.

Strieber, W. *Communion.* New York: Avon, 1987.

Sullivan, W. *We Are Not Alone* (Revised Edition). New York: Plume, 1994.
Williams, H. *Whale Nation*. New York: Harmony Books, 1994.
Wu, C. "Sometimes a Bigger Brain Isn't Better." *Science News*, 148(8) (August 1995): 116.

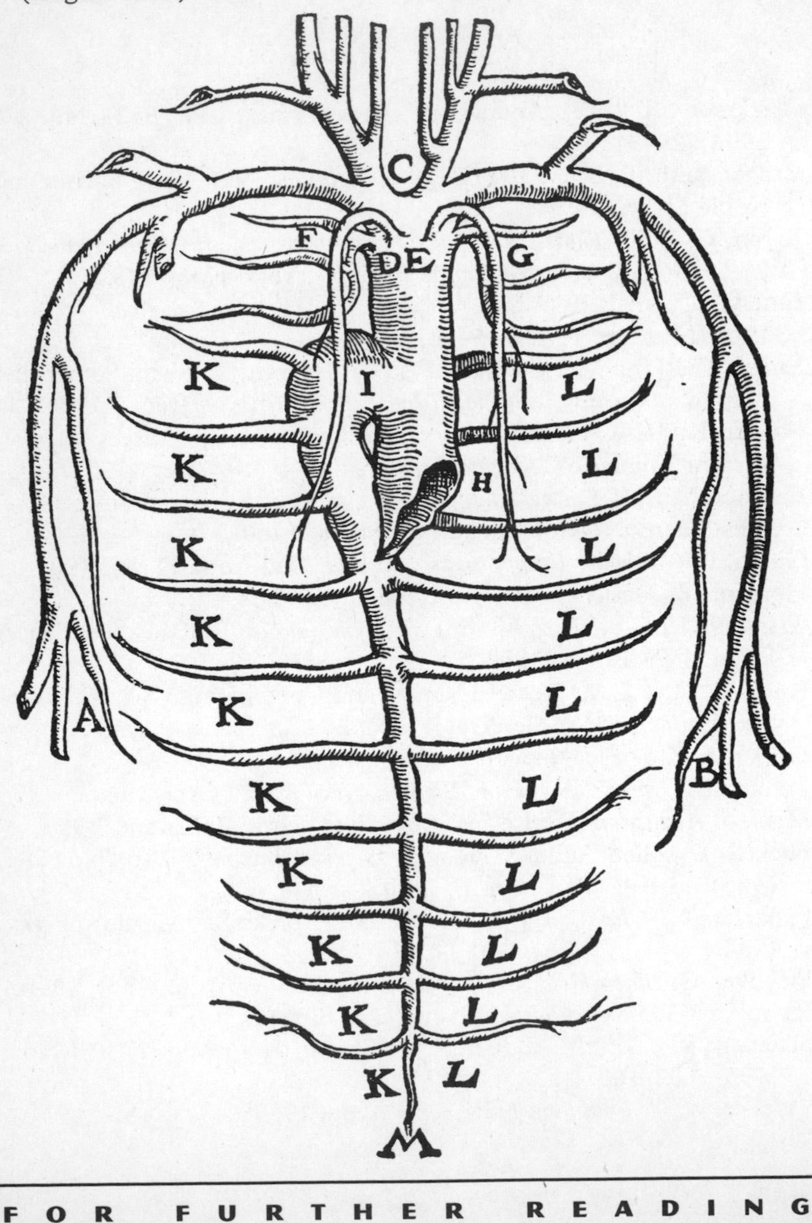

ATTENTION ALBUQUERQUE: CEASE
TRANSMISSION. REPEAT.

CEASE TRANSMISSION.
NATIONAL SECURITY ITEM.

DO NOT TRANSMIT.

STAND BY . . .

Alleged warning to an Albuquerque radio station
reporting on the Roswell, New Mexico, crash of an unidentified
flying object on July 2, 1947

THE MISSED MILLENNIUM

For ten of your terrestrial years we observed you as peers,
walking and talking among you,
fathoming the human condition
as best we could—what it means to be ephemeral,
to feel the flesh, to apprehend beginnings and endings,
to feel the flesh, to welcome insights like flashes
in the darkness, the darkness that is terra cognita
to the benighted, to feel the flesh—
and for the decade-long duration of our experiment
we nullified your awareness of the passage of time.
For you there was only the eternal now moment,
the augenblick that alone among you
Meister Eckhart seemed to understand.
For ten years we sowed your fertile soil
with gratuitous graces and subliminal messages
and watched you grow, then vanished back to
the bay of dust
and the keep of old forgotten dreams and
we snapped our fingers
and you all woke up. Bed-wetters, the lot of you.

So you've missed your new millennium,
it's ten years behind you
and the world didn't come to an end. But rest assured
we can arrange for you to arrange that for you.
Don't push your luck and you'll be fine.

-Keith Allen Daniels

ABOUT THE AUTHOR

Clifford A. Pickover received his Ph.D. from Yale University's Department of Molecular Biophysics and Biochemistry. He graduated first in his class from Franklin and Marshall College, after completing the four-year undergraduate program in three years. He is author of the popular books *Keys to Infinity* (1995) and *Black Holes: A Traveler's Guide* (1996), both published by Wiley, and *The Loom of God,* published by Plenum (1997).

He is also author of *Chaos in Wonderland: Visual Adventures in a Fractal World* (1994), *Mazes for the Mind: Computers and the Unexpected* (1992), *Computers and the Imagination* (1991), and *Computers, Pattern, Chaos, and Beauty* (1990), all published by St. Martin's Press. He has also written over 200 articles concerning topics in science, art, and mathematics.

He is also coauthor, with Piers Anthony, of the forthcomming science-fiction novel *Spider Legs*.

Pickover is currently lead columnist for the brain-boggler column in *Discover* magazine, an associate editor for the scientific journals *Computers and Graphics, Computers in Physics,* and *Theta Mathematics Journal,* and is an editorial board member for *Speculations in Science and Technology, Idealistic Studies, Leonardo,* and *YLEM.*

He has been a guest editor for several scientific journals. As editor of *The Pattern Book: Fractals, Art, and Nature* (World Scientific, 1995), *Visions of the Future: Art, Technology, and Computing in the Next Century* (St. Martin's Press, 1993), *Future Health* (St. Martin's Press, 1995), *Fractal Horizons* (St. Martin's Press, 1995), and *Visualizing Biological Information* (World Scientific, 1995), and coeditor of the books *Spiral Symmetry* (World Scientific, 1992) and *Frontiers in*

Scientific Visualization (Wiley, 1994), Dr. Pickover's primary interest is in finding new ways to continually expand creativity by melding art, science, and other seemingly disparate areas of human endeavor.

The Los Angeles Times recently proclaimed, "Pickover has published nearly a book a year in which he stretches the limits of computers, art, and thought."

Pickover received first prize in the Institute of Physics' "Beauty of Physics Photographic Competition." His computer graphics have been featured on the covers of many popular magazines, and his research has recently received considerable attention by the press—including CNN's "Science and Technology Week," *The Discovery Channel*, *Science News*, *The Washington Post*, *Wired*, and *The Christian Science Monitor*, as well as in international exhibitions and museums. *OMNI* magazine recently described him as "Van Leeuwenhoek's twentieth century equivalent." *Scientific American* featured his graphic work several times calling it "strange and beautiful, stunningly realistic." Pickover has received U.S. Patent 5,095,302 for a 3-D computer mouse and 5,564,004 for strange computer icons.

Dr. Pickover's hobbies include the practice of Ch'ang-Shih Tai-Chi Ch'uan (a form of martial arts) and Shaolin Kung Fu, raising golden and green severums (large tropical fish found in the central Amazon basin), and piano playing (mostly jazz).

He can be reached at P.O. Box 549, Millwood, New York 10546-0549 USA (Web site: http://sprott.physics.wisc.edu/pickover/home.htm).